PROTOCOLS FOR THE JUSTIFICATION OF RISK FROM RESIDUAL CONTAMINATION

CBRNE Ltd

PUBLISHED BY **CBRNE LTD**

i

Title:	D5.6
Date:	August 27, 2014
Author(s):	Nigel Hale, Dominic Kelly, John Astbury CBRNE Ltd (CBE), Jamie Braybrook

This project has received funding from the European Community's Seventh Framework Programme. The views expressed in this document are purely those of the writer and may not in any circumstances be regarded as stating an official position of the European Community.

Front Cover Design by: Carolyn Smith BA (Hons) Ind Des MFA - CBRNE Ltd Design Director

Contents

1. Executive Summary

An attack on a building or location using non conventional weapons will almost certainly lead to a requirement to undertake some form of decontamination of that site. This simple observation immediately raises the question "what decontamination standards and techniques should be used?". These questions are important and they have been the subject of much technical research – often under the banner of "how clean is clean" and "what is reasonable in terms of clean-up costs and decontamination measures". There is mounting evidence, however, that they cannot be considered separately from the related questions of "how should the standards be selected and applied?" and "who should be involved in that selection?". In the absence of consideration of these latter questions an official declaration that a building is safe for reoccupation is just that and it may be meaningless if the occupants and other stakeholders do not perceive it as safe. Conversely, the expert appraisal of a building as not yet safe for occupation may be out of step with the ideas of building owners and users who are eager to re-occupy a structure or part of it, for commercial reasons. For example, taking a contaminated site back to its original state may not be enough for some stakeholders and it may prove difficult to determine what the "original state" was for decontamination purposes. It may also be the case that insurers will only pay for what they consider to be "reasonable" decontamination measures; invariably it is the legal process that determines the definition of "reasonableness" in any particular case as reasonableness is not an absolute characteristic. The acceptable level of decontamination among stakeholders is not, therefore, a straightforward process.

The following document has two purposes,

> To present the results of work which has been undertaken, as part of Work Package 5 of project PRACTICE, to identify the factors that influence stakeholder perception of risk from contamination and to identify the additional information and assistance that should be gathered and presented to achieve acceptance of decontamination effectiveness. This is presented in Part A of the document.

> To support the PRACTICE goal to provide a Toolbox which will furnish the EU Commission and Member States (MS) with a flexible and integrated system for coordinated response to CBRN terrorist attacks in order to improve their chances of achieving effective and acceptable decontamination strategies. For that Toolbox, this document provides a series of high level Protocols which may be tailored by those who wish to adopt them to procedures and practices specific to their needs, These are presented in Part B of this document.

The document shows that the Source→Exposure→Pathway model, often used in environmental impact assessment, is a useful way to consider the various aspects of decontamination. It also shows that in normal practice (especially within the nuclear industry), decontamination projects encompass activities which do not necessarily remove contaminants but also those which interrupt these Source→Exposure→Pathways.

A review of previous incidents (see Section 5) involving contamination or similar concerns underpins the conclusions of this report that successful management of a decontamination programme following an incident and obtaining Stakeholder agreement is not a straightforward matter in that what is done both before and during the various phases of a decontamination exercise can shape the way that Stakeholders perceive risks. It can also affect the degree of decontamination and sampling that is required to obtain consent. The differences between an Organisation seeking to decontaminate something following an occurrence which might be considered to lie within its normal sphere of operations and one dealing with an abnormal incident are also highlighted. It is shown that that the assumptions of normal decontamination practice do not apply in the post incident situation; principally because of uncertainties and time pressures.

Much previous research has been presented in the matter of Stakeholder Management; this has been used to show the importance of the adoption of a formal Stakeholder Management Programme and procedures. It has also been shown that Stakeholders may be placed into one or more of three groups, namely the Organisation – representing the entity or organisation who has suffered the primary impact of the contamination incident and who is ultimately responsible for undertaking the decontamination (or at least paying for it), an Official Group – representing those with a formal mandate to make decisions (like Government bodies and Local Authorities/Councils etc) and finally a Public Group – representing those who are affected by the operations of the Organisation but who do not normally fall within either of the other two groups. There are overlaps between these group definitions which an Organisation must seek to understand.

Similarly much work has been presented in the matter of risk perception; this has been used to show how risk perception varies between stakeholder groups and how it is important for an Organisation to understand and manage this aspect of a decontamination project. A set of typical dimensions of risk perception is presented.

A model of Stakeholder dynamics – in terms of how Stakeholder objectives and goals can change as a project proceeds – is presented along with a potential model for the management of decontamination projects which implicitly includes the needs of Stakeholders. The value of an approach including aspects of Adaptive Management and Multi Criteria Decision Analysis is presented.

Ultimately, in Part B, a high level framework for six Protocols is presented. These cover Understanding Insurance, Stakeholder Identification and Analysis, Stakeholder Management, Stakeholder Communication, Dynamic Risk Management and finally the Response Phase Action Plan.

Given the potential breadth of Organisations (in terms of size and interests) that may wish to adopt plans such as those promoted here, it is not possible to provide a one-size fits all approach. The Protocols are therefore, specified at a high level. The protocols are not intended to be prescriptive but rather to highlight the key themes that need to be pursued by an Organisation in order that they might improve the likelihood of achieving a decontamination exercise that meets with both scientific and broader societal/stakeholder approval.

Where appropriate, the Protocols have been linked to existing European management standards such as those presented in EN ISO9001, EN ISO14001 and the European Framework for Quality Management. In this way it is hoped that Organisations will be able to implement the Protocols within existing frameworks of procedures and processes.

The target audience for this report is a semi-technical readership, perhaps consisting of those with responsibility for buildings, land or other assets that may become contaminated, who wish to make arrangements that will assist them in recovery from such incidents should they happen. The document is to be considered as a primer for such interested parties and as such it provides an insight into the subject matter and references to further studies that the reader may wish to consult.

Part A of this report presents an overview of the subjects of contamination, decontamination and recovery from a contamination incident, including some example case studies. Part B presents the resulting Protocols. The reader is recommended to read Part A so that the need for and context of the Protocols in Part B is better understood and appreciated. Some of the Protocols relate to preparedness (i.e. they should be implemented prior to any incident occurring) and some of them relate to processes and procedures that are more applicable following the occurrence of an incident.

This report is related to PRACTICE deliverable D5.12 "Targeted Action Plans" which provides guidance to Organisations who need to appoint and manage remediation / decontamination works following an incident of the type considered here. D5.12 deals with the practical, commercial and project management related issues rather than the justification issues addressed here.

Before attempting to implement any of the Protocols the question that an Organisation must answer is "how much risk am I prepared to take and how much effort am I prepared to take now to try to mitigate the remaining risk?"

2. Glossary

CBRNE	Chemical, Biological, Radiological, Nuclear, Explosive
Contamination	The addition to a substance or material of a second substance or compound (the Contaminant) – usually in much smaller quantities. In this report the Contaminants are such as to cause a nuisance or a health hazard and they are either C, B, R or N materials.
Decontamination	The process or removing contaminants or mitigating the hazard presented by them.
DRA	Dynamic Risk Assessment
EU	European Union
Hazard	Something with the potential to cause harm
MCDA	Multi-Criteria Decision Analysis
PRACTICE	Preparedness and Resilience against CBRN Terrorism using Integrated Concepts and Equipment
Residual Contamination	The contamination remaining at the scene of an incident after the recovery phase
Risk	A measure of the harm that could be caused by a hazard under specific circumstances.
Stakeholder	A person, body or organisation that has an effect upon, or is affected by, the operations of an Organisation.
Organisation	A business, building owner or other party whose assets have been affected by a contamination incident and who ultimately is responsible for ensuring that that contamination is appropriately addressed.
Public Group	Those stakeholders affected by an incident that are not members of either the Official Group or the Organisation (see separate entries).
Official Group	The Official group represents those with a mandate and authority granted to them under legislation or other national or local government arrangements.

PART A: A review of factors affecting acceptability of Risk from Contamination

1.   Introduction to Part A

The following sections of Part A provide an introduction to the Security Cycle presented further in PRACTICE Deliverable D3.1. Section 3 provides an overview of the concept of contamination from Chemical Biological, Radiological and Nuclear materials (CBRN). The standard approach to decontamination – during normal operations – is introduced and discussed in Section 4. Following this and some discussion of previous incidents involving CBRN contamination and a very brief précis of insurance arrangements in Europe, the applicability of these normal decontamination practices to post incident situations is presented and analysed. Section 7 pulls together all of the previous sections to derive the Protocols presented in Part B. In particular the need for Stakeholder Management and the recognition of the dynamic nature of stakeholder group membership and interests is discussed.

2.   Incident Response and Recovery

Before discussing the main topics of decontamination and what lessons have been learnt from previous incidents, it is useful to set out the phases of an incident and recovery from its effects and to show where this report sits in those phases.

Figure 1 Security Cycle

Taking the process outlined in Figure 1; during the Threat Assessment, Prevention and Preparedness phases an Organisation is going about its normal day to day business or activities; its normal business practices and management procedure apply. Preparing regular risk assessments and planning mitigation strategies bridges the gap between normal operation and when the Response phase is triggered. The purpose of the Recovery Phase, which normally does not involve emergency services, is the re-instatement of the Organisation's business activities or recovery of its assets;

The analysis presented in the remainder of this report shows that the justification of risk from contamination involves all phases in the diagram

3. CBRN Contamination

In its simplest form contamination is the addition of a substance to another substance that renders it unfit for its intended purpose. This simple statement can also apply to the contamination of people and property. Use of the word 'contaminated' gives no indication of the extent of the contamination or the degree of its impact.

Figure 2 shows the various ways in which contamination occurs when a substance A is added to the normal composition of substance B so that the substance is now a combination of both A and B. Substance A is said to be contaminating substance B or to be a contaminant of B.

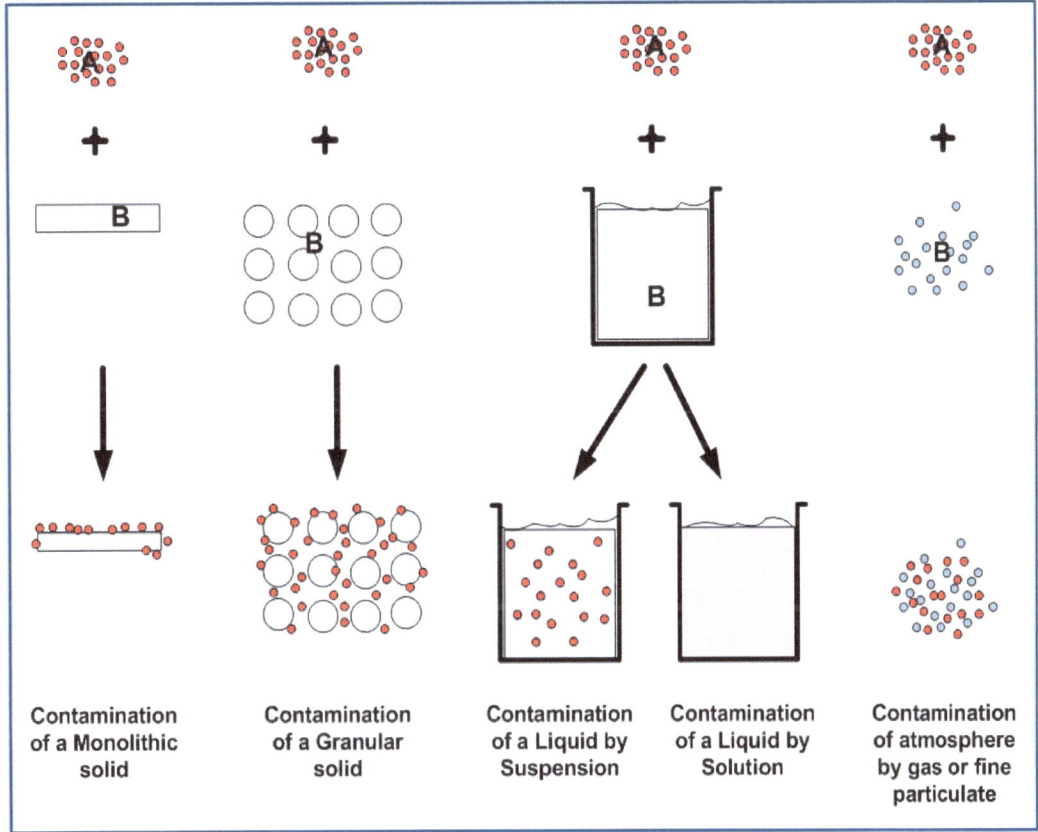

Figure 2: The Different Processes of Contamination of a Substance B, with Substance A

Notes to Figure 2: If B is a solid monolithic material (e.g. a steel object) then the contaminant may simply sit upon its surface or it may have formed a physical or chemical bond with it (depending on the material properties of A and B), but it will not normally have penetrated the bulk of the material to any appreciable depth . If, however, B is a granular material (e.g. soil, concrete, sand etc) then the contaminant may become inherently mixed in with the bulk of the material (depending on the method by which the contaminant is transported) and may become physically or chemically bonded to it. If B is a liquid then the contaminant may either become suspended in the liquid (i.e. it may retain its separate physical nature) or it may dissolve in the liquid (i.e. it may breakdown into smaller chemical parts and become intrinsically mixed with the liquid).

The ways in which substances can become contaminated varies, depending upon whether they are solids, liquids, gases or mixtures of these; this is noted in further detail in the notes to .

Each of the different mechanisms of contamination presents different challenges with respect to decontamination. Furthermore, real life objects clearly have much more complex geometries and structures than those shown in the simple diagramme. These complexities mean that gaining access to parts of objects that may have become contaminated may prove very difficult. Where the contaminant is not physically or chemically bound to another object or surface then it can be disturbed and may spread further. This is known as migration.

The following sections give a very brief overview of the different types of contamination that can occur following a CBRN attack or an incident. PRACTICE Deliverable D2.1 (Scenario template, existing CBRN scenarios and historical incidents), provides more detailed definitions of the different types of materials that may be involved in CBRN contamination incidents.

The routes through which contamination may affect people, animals and objects are shown in Figure 3 which is discussed in Section 4.

3.1   Chemical Contamination

All substances are composed of chemicals in the final analysis. In the acronym CBR, Chemical implies materials which are not living (biological) and which do not present radiological hazards. As such it represents materials in all phases, i.e. solid, liquid and gaseous. They cause harm by, for example, being toxic, poisonous, harmful to the respiratory tract or by affecting the nervous system. For Chemical contaminants to cause harm they must be ingested, inhaled, injected into or come into contact with y humans.

3.2   Biological Contamination

The word Biological implies that the contaminant is a living organism such as a bacteria or a virus, but it also includes toxins (chemicals) produced by organisms. They may be in solid, albeit very small, or they can be liquid. Although they may be transferred via air movement they do not constitute true gases.

For Biological contaminants to cause harm they must be ingested, inhaled, injected or touched by humans.

3.3   Radiological and Nuclear Contamination

Radiological materials are those that undergo what is known as radioactive decay during which they emit radiation which may be harmful. Radioactive materials may be solids liquids or gases; they do not generally change their physical form or chemical properties following decay[1]. Some radioactive materials may also be hazardous in their own right, i.e. they would be harmful even if they were not radioactive, for example Polonium-210 which was used to poison Alexander Litvinenko is toxic when ingested regardless of whether it is radioactive or not. The radioactive

---

[1] Some radioactive materials, usually those formed in nuclear reactors or following atomic detonations, undergo a type of decay in which they change into other materials. This is known as fission.

forms of a material are known as isotopes. Some materials have more than one isotope (Polonium 210, for example, is one of the isotopes of Polonium, but there are others).

In contrast to C and B materials, whilst radiological contaminants can also be harmful if ingested, inhaled, injected or touched they can also cause harm at a distance due to the ability of the emitted radiation to travel from the source material to people and cause harm there. The distance that can be travelled and the amount of harm caused depend upon the type of radiation and how much of it there is. Note that radiation itself is not a contaminant, it is the source material that emits the radiation which is the contaminant.

Nuclear contamination refers to radiological contamination produced by an atomic detonation. In this respect it is just another form of radioactive contamination, although the materials produced are generally extremely hazardous and concentrated and are associated with very large amounts of physical energy and damage[2].

4. The normal approach to decontamination

The following sections discuss the normal approaches to decontamination and show how some of the assumptions upon which they are based do not apply to terrorist incidents.

4.1 Decontamination Types

To understand the effectiveness of decontamination and the hazards presented by it, it is useful to consider the source-pathway-exposure model as presented in Figure 3. The figure shows how, once some hazardous material has been released it may be dispersed or transported and then may ultimately lead to exposure via inhalation and ingestion, for example. The degree to which the various modes apply will vary from incident to incident . At the right hand side of the figure the various dispersion and transportation modes result in contamination of water, food etc.

---

[2] Atomic detonations can create additional radioactive material via a process known as activation.

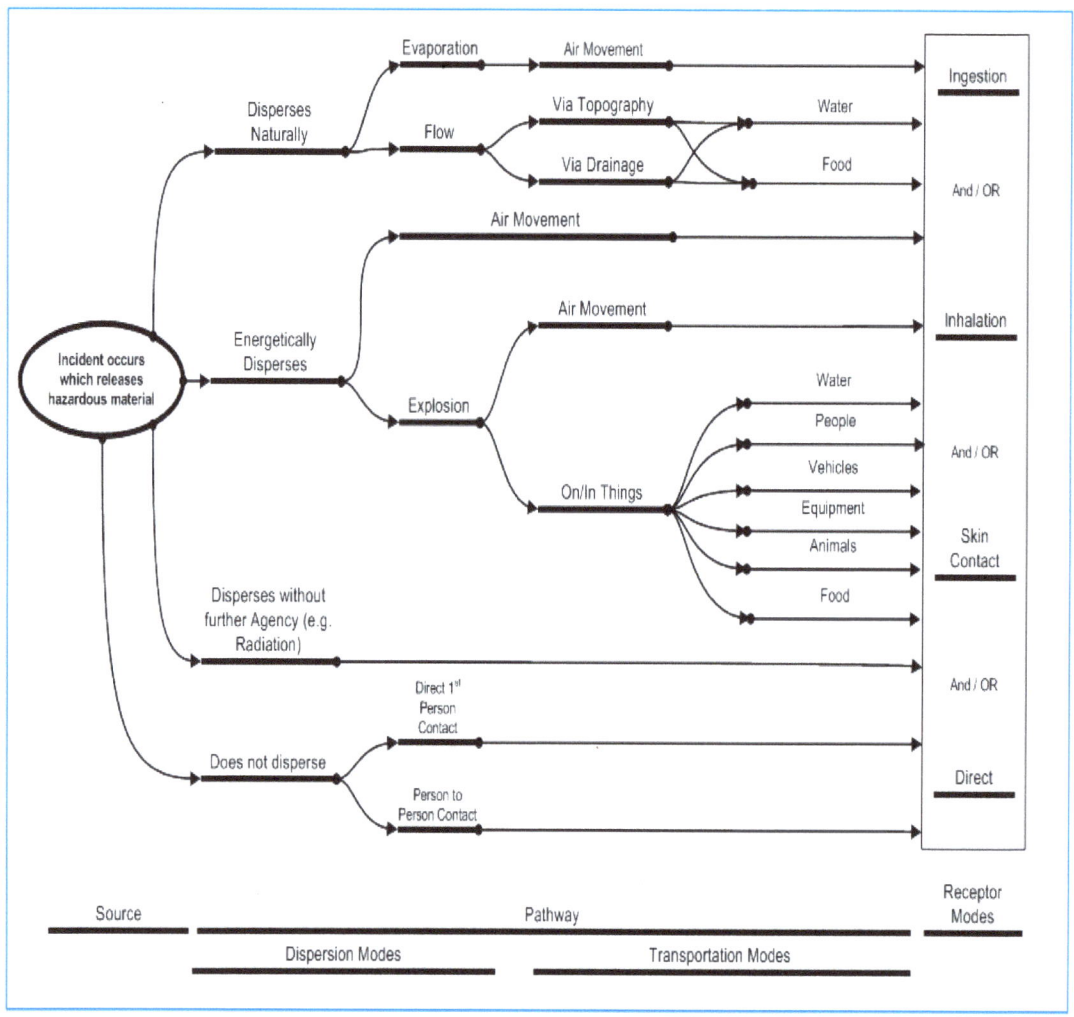

Figure 3: Source→ Exposure→ Pathway Model

12

With respect to the control of contamination, examination of shows that reduction of the hazard can be effected by intervening at any stage, for example:

a. by removing the source material itself,

b. by preventing the dispersion modes,

c. by interfering with the transportation modes,

d. by intervening at the receptor end of the chain.

Although decontamination in its strictest sense is clearly applied at point "a", in general practice decontamination projects also include activities of types "b" and "c" [3] as these prevent further spread of contamination and also reduce risk.

Thus, in its broadest sense, decontamination can include the activities shown in Table 1 [Esposito et al, NISD, AF Colenco et al]. .

Table 1: Types of Decontamination Activities

	Type	Description
A	Complete Removal	Removal of either the contamination itself – e.g. by extraction of contaminated material (and sometimes some of the material on which it sits/is mixed with – e.g. the removal of topsoil) or wholesale removal of the contaminated material and the material on which it sits/is mixed with.
B	Partial Removal	Removal of some of the contaminated material to the extent that any remaining contamination does not present any concern for safety – e.g. because it is either at such low levels that it is considered to be safe or because it is "fixed" (see immobilisation below) and hence can not be transported to present an exposure route.
C	Neutralisation	Conversion of the material to a less hazardous or non hazardous form – e.g. by the application of a biocide to remove a biological contaminant or application of a neutralising agent to reduce the hazards from an acidic material.
D	Immobilisation / Containment	Fixing of hazardous material in place such that it can not be dispersed or transported to people and cause harm – e.g. by the application of a fixative (paint, varnish, epoxy resin etc) or by encapsulation (in concrete, plastic etc).
E	Pre-cautionary techniques	The treatment of material which may reasonably be expected to be contaminated but for which reasonably practicable methods of identification or removal / neutralisation do not exist. Precautionary removal, neutralisation and immobilisation are included in this definition.
F	Combinations of some/ all of the above	See above.

Note to Table 1: B, C and D and some examples of A necessitate methods of detection that are adequate to demonstrate that remaining material is contamination free or that remaining contamination is below some agreed "safe" level.

---

[3] Intervention at point "d" represents the provision of personal protective equipment and/or respiratory protective equipment and as such does not represent a long term strategy or decontamination.

4.2 Decontamination Techniques

Table 1 has shown the different types of decontamination activities and has provided some indications of decontamination techniques. Table 2, however provides further details of some techniques commonly used in industry for types A, B, C and D (as defined in) – to provide background to the discussion that follows in the remainder of the report. The examples provided are not exhaustive or comprehensive; each instance of contamination requires careful consideration to identify the appropriate decontamination strategy or strategies. The table demonstrates some of the issues that are associated with decontamination and some of the secondary hazards that may arise. Some of these are key to understanding the 'whole' risk presented by contamination and the perception of risk as discussed in later sections of this report.

Table 2: Examples of Decontamination Techniques

	Type	Description and Comments
A	Complete Removal	This can range from simple cleaning techniques using detergents and solvents or vacuum equipment to aggressive techniques like surface scabbling to remove the contaminated surface and the contaminants, to simple excavation of material (e.g. top-soil) or complete removal of contaminated items of equipment. Some of these techniques carry additional risks associated with the generation of secondary airborne hazards – e.g. surface scabbling – such that they require careful protection and isolation – e.g. ventilated and filtered containment structures. Clearly, performing these activities can also present a hazard to the operatives undertaking them and the risks arising from these operations need to be carefully weighed against the long term risk reduction that may result from them. These techniques may also generate significant volumes of contaminated material - which will require either safe disposal or further treatment off-site – and may change the bulk physical form of the contamination – e.g. from a solid or a dust to a liquid suspension – which may bring its own problems.
B	Partial Removal	Partial removal entails the use of the techniques listed against A but only to the extent that some of the contamination is removed rather than all of it. The choice of partial rather than complete removal may be due to practicable or financial constraints and may therefore embody some degree of risk optimisation.
C	Neutralis-ation	Neutralisation usually involves the application of an agent of some form to the contaminated materials or exposure of them to some form of external agent. Examples include the application of biocides to biological contamination (e.g. hydrogen peroxide, chlorine dioxide, bleach solutions and proprietary biocides), application of other chemicals to initiate a chemical conversion to less hazardous substances – (e.g. the application of weak bleach solutions to areas contaminated with sulphur mustard) and exposure to UV light to kill of biological organisms and/or initiate chemical breakdown. It is clear from the above that many of these techniques involve the use of chemicals which themselves may be hazardous and which require careful application and control. As for the techniques at A, they may also generate large volumes of hazardous or difficult wastes, albeit less hazardous than the initial contaminants. Neutralisation is not relevant to radiological hazards.
D	Immobilis-ation / Contain-ment	Although immobilisation does not itself remove the contaminants it can provide a cost effective and appropriate response to certain contaminants as it can provide protection against direct contact with contaminants and against exposure to airborne hazards, which from can be seen to reduce the risk of exposure. Typical examples include the application of epoxy based paints or other non water based paints, application of spray coatings (usually polymer based) or bulk sealing of large items by encapsulation in a solid matrix such as concrete or resin. At the simplest of levels these techniques could include simple containment within doubly sealed polythene bags. These techniques are generally considered to be of a low hazard but non permanent. In some instances, such as radioactive contamination, their limited lifetime may, however, be sufficient to protect against exposure until such time as the hazardous material has naturally decayed away in any event. Typically, where these techniques are used they are accompanied by rigorous monitoring regimes to ensure that the immobilisation remains effective. Similarly, the designated use of areas containing contained or immobilised materials may be changed to protect them from physical damage –e.g. by the exclusion of vehicles.

4.3   Decontamination Sequence

Decontamination during the normal operations of an Organisation, is normally a linear series of activities such as shown in Figure 4 [4] [IAEA 2009, Whicker J.J et al].

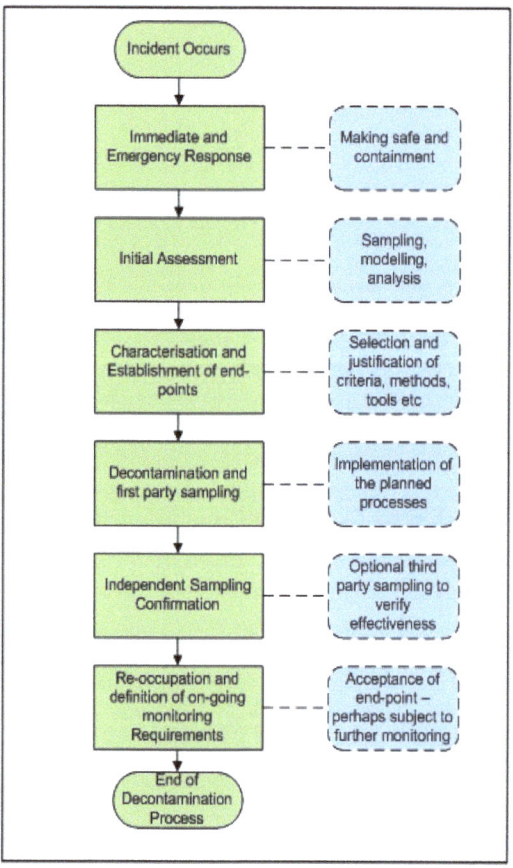

Figure 4: Linear Approach to Decontamination

This approach is widely used in industry and has proven to be effective. Consideration of Figure 4 however, indicates that application of this normal practice requires a number of elements as shown in Box 1. The relevance of these to post incident conditions is discussed later in this document.

---

[4] Some non-linearity is present in as much as some areas of a building may need further treatment after sampling but this normally only requires further application of the same techniques and is linear in this respect.

```
┌──────────────────────────────────────────────────────┐
│          Box 1: The Predicates of Normal Decontamination        │
│                        Practice                        │
│                                                        │
│   •      the contaminants and their behaviour being known,       │
│                                                        │
│   •      the availability of proven decontamination techniques,  │
│                                                        │
│   •      the availability of suitably trained operatives,        │
│                                                        │
│   •      the suitability of existing facilities and identified waste │
│          routes,                                       │
│                                                        │
│   •      the existence of an existing and suitable hierarchy of  │
│          management,                                   │
│                                                        │
│   •      the existence of previously accepted decontamination    │
│          standards,                                    │
│                                                        │
│   •      the existence of budgets and clear imperatives (in-     │
│          house financial considerations) to quickly achieve the  │
│          decontamination.                              │
└──────────────────────────────────────────────────────┘
```

4.4   Decontamination Proof

In order to demonstrate that a decontamination programme has been successful it is clearly necessary to obtain some form of evidence based data (samples) for comparison against a chosen standard (the end-points referred to in Figure 4). The selection of an appropriate standard will consider the following factors:

i)   The potential exposure pathways of concern (see Figure 3)
        This will depend upon, for example, whether the exposure route of concern is inhalation (in which case a measure of the airborne concentration and likelihood of the material becoming airborne is appropriate) or ingestion (in which case the potential for the contamination to spread to the food chain might be appropriate).

ii)   The nature of the potentially exposed population
        For example, a different standard may be selected for adults, children or workers.

iii)   The degree of certainty that exists about whether or not items may have become contaminated or not[5].

For example, an item may have been located in an area for which there has been no identified route for the contamination to reach or an area for which there is no evidence for the contamination having reached. In both of these cases, contamination is not expected but the latter may require more detailed proof and hence more rigorous sampling than the former. See Figure 5.

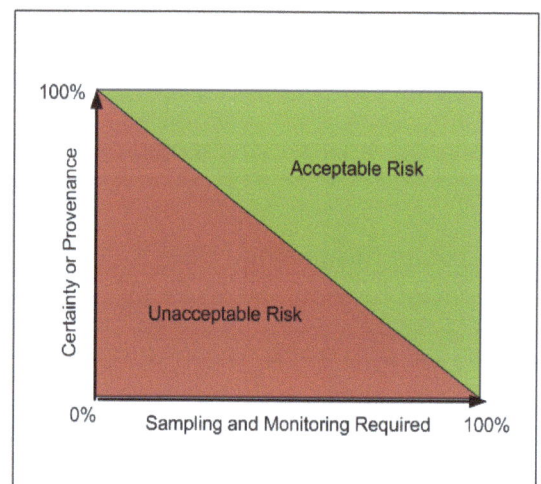

Figure 5: Risk Acceptability, Monitoring Requirements and Provenance[6]

The selection of appropriate techniques for obtaining sample data will depend upon factors such as;

i)   The potential exposure pathways

For example, a contaminant for which the exposure pathway of concern is inhalation may necessitate obtaining air samples whilst one that causes harm through ingestion may require analysis of food samples.

ii)   The likely extent of the contamination

For example, there may be good physical arguments to support a claim that contamination is only likely to be found in Area A, but not in Area B. In this case Area A may be sampled extensively and Area B less so.

iii)   The efficiency of the method for obtaining samples / data

---

[5] There are sometimes instances where it is not possible to specify any monitoring and sampling methodology that will provide sufficient assurance that something has been decontaminated sufficiently. In these instances there is no choice but to presume the presence of contamination and to dispose of or destroy the whole item. See E in Table 1.

[6] Adapted from [CEWG]

Some materials are very difficult to sample or may require large samples and long analysis times in order to be able to obtain accurate measurement. Others may have areas/volumes that are difficult to access.

iv) Cost and Programme

There may be multiple techniques available for obtaining samples or direct measurements but some of these may not be compatible with or justifiable against budgetary constraints or other pressures on timescales and programme.

v) Some contaminants require costly and time consuming analysis
Some sample analyses may be expensive to undertake, especially if complex laboratory analysis is required.

vi) Statistical Confidence required
This relates to the risk associated with the residual contamination and the permissible margin for error – noting that it may not always be possible to undertake 100% sampling of all (potentially) contaminated areas.

It is clear from the above that, in a similar way to the choice of decontamination approach varying with contaminated item, the choice of decontamination standard and sampling technique will also vary.

5. Reviews of Previous Incidents

A comprehensive listing of terrorist related and accidental incidents involving CBRNE material is presented in deliverable D2.1 for the PRACTICE project [FFI]. Further histories are available on the internet.

Brief summaries are presented in Sections 5.1 to 5.9 for several contamination incidents which have occurred in recent times. A summary of the key findings is presented in Section 5.10.

Although Deliverable D2.1 presents many examples of incidents not all of them led to contamination which required removal – many of them lead to the release of material which naturally dispersed or degraded without intervention. Due to the scarcity of such incidents in Europe some of the case studies are taken from America. Notwithstanding the differences in political arrangements, national organizational structures and ethnicity, these cases are considered to be directly applicable to the current topic. In any event this document deals more with issues of perception and involvement than those of scientific exactness; the author believes that people are fundamentally the same regardless of their country of birth and origin – in any event America's population is clearly multi-ethnic and multi-religious – but the effects of cultural differences are recognised in the literature, albeit that they are the subject of some debate.

The purpose of presenting these case studies is to highlight some of the factors that are at play following an incident and to provide indicators to some of the areas that warrant further investigation for this report. It is not intended to be a comprehensive listing of previous contamination related incidents.

The incidents that are reviewed are;

- The deliberate contamination of public and government sites by Anthrax spores in 2001– now commonly referred to as the Amerithrax incident.

- The accidental contamination of the Binghamtom offices in New York in 1981, referred to by some as the SOB incident.

- The accidental contamination of numerous premises as a side effect of the Polonium 210 poisoning of Alexander Litvinenko in London in 2006

- The Ammonium Nitrate Explosion in Toulouse in 2001.

- The orphans source incident in Turkey in 1998.

- Decontamination of an Illegal Chemical Laboratory Site in the UK in 2008.

- The Smailholm Anthrax Incident in Scotland in 2007.

- The Tokyo Metro Sarin Attacks in 1995.

- Decontamination of an Illegal Drug Laboratory in the UK

5.1 Amerithrax

In 2001 there were several incidents in which Bacillus anthracis was posted to various public and government sites in the United States. This lead to several deaths from inhalational anthrax and to more than 20 cases of inhalational or cutaneous anthrax [NAS]. The contamination affected not only those who opened the letters or who were at the site where the letters were opened but also at locations along the route through which the letters passed. Decontamination was therefore required at numerous locations.

In the aftermath of the incident decisions had to be made about which sites required decontamination and by what method. Moreover, there were uncertainties over what the decontamination end-points should be used and how they should be measured. i.e. how "clean" the buildings had to be for reoccupation.

The results of consultations that took part during the clean-up exercise revealed views that ranged from support for the authorities to complete "distrust and hostility" [NAS].

The clean-up exercise following the attacks was technically challenging in so far as such an exercise had not been attempted before for Anthrax contamination of such a large scale in a public building[7] and challenging in respect of the Public perception of the risks arising from the

---

[7] Previous experience did exist of decontamination of Gruinard island in the UK, known as "anthrax island" after it was deliberately contaminated by the UK military in the 1940s. [Pearson, 1990]. The decontamination techniques used and the isolation of the island from people mean that the incident was of little relevance to Amerithrax.

attack. Thus the works were associated with a degree of uncertainty and change as scientific knowledge and experience developed. One particular area of debate was that concerning the decontamination standards to be used and what was acceptable to the building owners and the public. There was also some debate about how samples should be obtained for testing against the standards.

In total NAS2006 presented 27 findings and an additional number of related recommendations. The four key findings and recommendations relevant to this study are presented below[8], the referenced document should be consulted for the full text of the findings and recommendations;

Finding 3-1

The determination of acceptable risk is a complex issue and the willingness to accept risk varies from person to person, from situation to situation, and from culture to culture. Managing risk also is complex: Different people have different ideas about how much responsibility the government or the owners and operators of public facilities and lands have to limit public exposure to risk. ...

Recommendation 3-1

Authorities should base their plans on lessons from the experiences of others who have dealt with decontamination issues in the broadest sense; they should not consider their charge a completely novel task.

Finding 3-2

If safety-related standards and protocols are devised and implemented behind closed doors, without the consent or input of affected and interested parties, those standards are likely to be questioned or rejected outright. Lack of transparency for policy decisions that directly affect public health— even in the context of a proclaimed national security interest— can severely erode public confidence. The establishment of a formal planning procedure that involves relevant stakeholders before an event should expedite the response and confer legitimacy for decisions made during and after decontamination.

Recommendation 3-2

Representatives of affected parties should be involved in risk management decision making, and they should participate in the technical discussions needed to make decisions. Engaging the people whose well-being is most at stake helps ensure their greater confidence in the outcome of risk-based decisions. Those who provide the technical information should be independent experts who are free of conflicts of interest, so that they can give the highest priority to protecting public health. Stakeholder involvement in risk assessment and management provides valuable returns: local knowledge that can contribute to a more robust definition of the danger, greater public confidence in scientific tools that support public policy, and more widespread acceptance of the legitimacy of the results.

Finding 4-1

Acceptability is not a technical concept. It is a values concept. It is, therefore, best constructed through an analytical and deliberative process that involves key stake- holders in a potentially harmful situation. Without trust, acceptability is difficult to achieve. Effective leadership in dangerous situations is based on openness and honesty, even when bad news must be conveyed. Transparency in decision making can contribute substantially to ensuring the acceptability of risk. Panic is rare in disasters, and it is an unhelpful idea for explaining how people respond to frightening situations and information. After the 2001 anthrax attacks, decision makers sometimes relied on assumptions that later proved unfounded; their subsequent actions resulted in significant problems with communicating the degree of risk involved to the stakeholders.

---

[8] Reproduced under Copyright Licence 2806521095678 from RightsLink - Copyright Clearance Center 222 Rosewood Drive Danvers, MA 01923 USA.

Recommendation 4-1

Risk managers who face potential contamination should assume that the problem could be worse than they initially think. In remediation projects, the public should be seen as an asset, not a liability, and information should be made available widely. Indeed, the public should participate actively in decision making in the aftermath of an attack. Following the lead of previous work by the National Academies, the committee recommends that an analytical deliberative process (i.e. a combination of analysis and deliberation) be used to determine appropriate approaches for cleanup.

Finding 4-2

Relevant data from the sites contaminated in 2001 were not shared with all necessary parties, partly because of the differing goals and objectives of law enforcement and public health agencies. Lack of data sharing can compromise health in the aftermath of a biological attack.

Recommendation 4-2

Agencies and organisations entrusted with data relevant to public health should make every effort to share this information. Cooperation is the key to decreasing public anxiety…and agreements…should be in place.

5.2 SOB – The accidental contamination of Offices in Binghamton, New York

The Binghamton Sate Office Building (SOB) in New York was the site of an electrical transformer fire involving the release of Poly Chlorinated Biphenyls (PCBs) and their combustion products in 1981. The smoke and vapours produced by the fire caused widespread contamination of the office buildings and the surrounding areas.

The incident was characterised by poor communication both within the organisations responsible for the clean-up and between them and the public and by the attempt to justify the risks on purely scientific grounds. Although the technical issues were important they neither posed the most important dilemmas facing the decision makers nor provided solutions to them [Carke, L]. Not least among the issues associated with the proposed technical approaches was a failure to communicate or realise the uncertainties associated with the toxicological data for PCBs.

The early response of the numerous organisations involved in the immediate aftermath of the event was only very loosely co-ordinated with few established rules to govern their behaviour. In particular, the normal processes and procedures for co-operation and communication between the official bodies involved did not survive the incident,; they were either inappropriate for the conditions after the incident or they were conveniently viewed as such by some of the parties who would otherwise have been drawn into the post incident negotiations.

The combined effect of these problems was that there was a lack of trust and perhaps respect between the various stakeholders involved; the title "Stakeholders" includes the owners/operators of the building, the relevant authorities and the wider public bodies.

During the immediate aftermath of the incident, those officials responsible for the decontamination tried to demonstrate that the risk to stakeholders was low by comparing the risks from the contaminants with those from other sources of PCBs (e.g. fluorescent lights and refrigerators) rather than trying to define the incremental risk from the contamination incident itself. Further outrage was caused by the apparent lack of appropriate safety controls that were applied to some of the initial decontamination staff – who were part of the normal janitorial staff and hence were not equipped (either in terms of equipment or training) to appropriately handle the contamination hazards.

Perhaps the most important aspect of this incident is that the building has gone through several clean-ups to date and although the State of New York has declared that the remaining contamination presents an acceptable level of exposure to those who work there it is not clear even now that levels were reached that all agreed were safe [Clarke, L]. As for the Amerithrax incident there was some concern from the stakeholders about how samples were obtained for comparison with the chosen decontamination standard.

One anonymous official involved in the incident summarised it as "The major problem is not cleaning the building, but convincing people that it is safe to re-enter. This thing is more of a public relations problem than a technical one. The science isn't hard – the political process, the public relations, the sociology, the psychology of the whole thing is the problem". [Clarke, L].

5.3   The Poisoning of Alexander Litvinenko

On 1 November 2006, Alexander Litvinenko a former officer of the Russian Federal Security service, who was living in the UK under asylum, suddenly fell ill. He was later found to have been poisoned by polonium-210. The incident was more significant for the fact that his attackers had been careless with respect to the degree of contamination that they caused – notably at the Millennium Hotel in London's Mayfair district but also at numerous other locations.

Acton et al (2007) provided a review of some of the issues associated with the incident. Among their conclusions is that the case created comparatively little public concern and that this was due primarily to the perception that it was a targeted assignation attempt against a single person - those who held this view perceived the potential risk to themselves as much lower than those who felt that the incident was somehow related to the broader terrorist threat - and also that a key message put out by the Health Protection Agency at the time of the incident had been picked up by the public: most people knew that if they had not been to one of the areas known to be contaminated with polonium-210 then the risk to their health was nil. Clearly, successful communication in this case did much to alleviate public anxiety.

Although the authors suggest caution in extrapolation from this event they conclude that "Nonetheless, the incident does serve as an example of how effective risk communication can help reassure the public."

An additional aspect of the incident, not within the scope of the review by Acton et al, with which the authors have direct experience, was the determination of final decontamination levels for use by the responsible authority in determining whether or not areas and assets were "safe for public use" and could therefore be released from their control. The level set was within the detection capability of monitoring instruments and as such did not ensure that items were free from contamination to below detectable levels but rather that they were only contaminated to levels that were considered to be scientifically 'safe' (i.e. they presented negligible and acceptable risk, as judged by the technical authorities). Whilst there is little doubt that the levels set were appropriate from a strictly scientific view of risk, there were instances where third parties expressed the view that these levels were unacceptable from the standpoint of them asking members of the public or visitors to subsequently use these decontaminated assets[9]. In these cases, the owners of the assets requested decontamination to the extent that polonium-210 was no longer detectable. The suggestion here is that the third party view of what was reasonably practicable differed from that of the responsible authority, not that the official decontamination standard was flawed. It is worth noting that the incident caused contamination of a wide variety of surfaces and that for some materials it was deemed by the authorities that demonstrably safe decontamination could not be demonstrated in the field – i.e. any such materials for which there was any reasonable suspicion of contamination were removed and disposed of as hazardous waste. In this case the official view and third party needs were aligned. Thus, although the incident supports the view expressed in NAS (2005) that "acceptability is not a technical concept, it is a values concept", these can coincide.

In some locations that needed decontamination, the transfer of information from the first responders to the building owners and Westminster City Council did not operate efficiently and this led to some additional costs and delays that could perhaps have been avoided.

It is also noteworthy that, in contrast to the incident at Binghamton, the science of detection, quantification and decontamination of this type of contamination could be considered to have already been mature at the outset of the incident.

A further aspect of this incident worthy of note is that there were clear lines of responsibility, accountability and authority set out from the early phases of the incident and adhered to throughout[10]. Notably, the responsibility for Gold Command was taken by Westminster City Council (WCC) with technical support being provided by the Health Protection Agency, Health Authorities and the like. In this instance, the need for the Local Authority (WCC) to assume Gold Command rather than a more traditional enforcement role was perhaps due to the widespread and numerous locations affected and their public nature. Nevertheless, there were issues associated with the funding of the decontamination and the establishment of financial accountability between the owners/operators of the contaminated assets and their insurers.

5.4   Toulouse – Ammonium Nitrate Explosion

On September 21st, 2001, at 10.17 am, about 300t of ammonium nitrate exploded in a warehouse of the AZF plant in Toulouse, France. This explosion caused the death of 30 people, caused injuries to 2,500 people, and damaged 27,000 houses and apartments; the total amount of the damage was assessed at €1.5 billion. The cause of the explosion is still unknown [Dorison].

When permission was sought to re-open the plant the issues that caused the most difficulty were far less technical than social and political; those with an economic interest were mostly in favour of

[9] The exact location of such premises is covered by commercial confidentiality and can not be revealed in this report.

[10] WCC ensured that their authority was respected by the use of Enforcement Notices which ensured that suspect premises could not be put back into public use without their prior approval.

reopening, albeit with increased safety conditions, but most local associations, especially those of victims and those in defence of the environment, were against, whatever the economic and social price to be paid. Local politicians were also divided.

Much of the plant has now been demolished and there is evidence of strong psychological trauma still in the local population.

Although this was not a contamination incident, its significant and catastrophic nature demonstrates how, at the extremes of consequences – instantaneous or near instantaneous death of victims – scientific and economic arguments are of little value when discussing acceptable conditions for reinstatement of facilities with public stakeholders. The case also provides evidence of how different social interests and concerns influenced the acceptability of the risk, as between those with personal economic interests and those concerned with the environment.

5.5   The orphan source accident in Istanbul, Turkey, 1998

A company specialising in transportation of radioactive material received three ^{60}Co sources from old cancer treatment equipment in 1993. The company applied for and received permission to export these sources to the United States, where the original supplier would dispose of them as radioactive waste. For unknown reasons, the sources were packed for shipping, but never actually sent. In 1998, two of the containers were removed from a storage facility in Ankara to a warehouse in Istanbul. When this warehouse was sold, the new owners found the containers and, not recognizing the symbol for radiation, sold them as scrap metal [FOI].

The shielding was subsequently stripped off the sources, leading to radiation exposure of the workers in the scrap metal facility. Three days later, ten people who had worked in the proximity started showing symptoms of radiation poisoning, six of them vomiting. The cause of the illness was not identified until four weeks later when a doctor suspecting radiation poisoning contacted the authorities. In the meantime, at least one source had been left unshielded in a residential area for several weeks. The clean-up of the facility proved difficult. Only one source was found, and it is unclear whether the second container had contained a source at all. The accident is reported in an IAEA report [IAEA, 2000].

The news media's coverage of the incident continued for several days and created much anxiety. Both the TAEK (Türkiye Atom Enerjisi Kurumu – Turkish Atomic Energy Authority) and the medical authorities had to deal with many inquiries from members of the public who were concerned about their health. The open public information policy of the TAEK administration reportedly helped to allay the public's concern. The psychological impact of the accident on the public was high, as may have been expected [IAEA, 2000].

5.6   Decontamination of an Illegal Chemical Laboratory Site, UK[11]

---

[11] The exact location of the premises is covered by commercial confidence and can not be revealed in this

25

Following the death of a retired scientist in 2008 his family discovered numerous chemicals at his house and in his rear garden. The house is located on a busy main road and is bounded on the remaining three sides by inhabited properties. Following initial investigations by the Local Authority they took control of and responsibility for the premises and contracted a specialist team of decontamination experts to undertake a detailed investigation and remediation project. Prior to any works being undertaken, the Local Authority representative undertook consultations with the neighbouring householders and businesses to advise them of their intentions to remediate the premises and to obtain their views on appropriate precautions to be undertaken. In this instance, the intention was to completely remove the chemicals and any other contaminants found (including some radioactive materials, asbestos sheeting and numerous chemicals), so that agreement on decontamination end-points was not an issue. Somewhat surprisingly, the neighbours' views did not place any restrictions on the works other than that they should be undertaken during normal daylight hours when they were predominantly absent from their properties in any event. One neighbour remained at their premises apparently unconcerned about the presence of 'men in white suits' working in his neighbour's garden. The immediate neighbours also offered their premises for use by the team as a staging area for the works. Once the works to clear the property had started, the extent of chemical and radiological waste at the site was found to be much greater than had previously been suspected but again, the Local Authority provided early communication of this to the neighbours and confirmation that the materials would still be wholly removed to below detectable levels.

The premises were decontaminated and the waste materials removed without incident and to the satisfaction of the Stakeholders. At one point during the remedial works the neighbours enquired of the decontamination team, who were at that time dressed in hazmat overalls and wearing half face respirators, if anything untoward had been found. When the team advised them that little of any significance had been found, the neighbour continued with their business apparently unconcerned.

Because of the potentially hazardous nature of some of the materials found at the site, the local Police and Fire services were also notified, but they too placed very few additional restrictions on the project[12].

This simple example provides good evidence for the importance of early stakeholder involvement, simple straightforward communication from a trusted source, clear lines of communication and examples of the sorts of straightforward expedient measures that may sometimes be sought from Stakeholders.

Since the neighbours had been living next to the hazardous materials for some time prior to the works being undertaken, albeit that they were not fully aware of the quantities of the materials, there is perhaps a suggestion that they had come to worry less about the risks associated with them and that perhaps their perception of the risk was lower than it might otherwise have been if they had suddenly become aware of the materials.

---

report.

[12] They only required regular updates on progress and assurance that the premises were secured against unauthorised access. The local Police arranged for regular visits during silent hours to ensure the continued security of the site.

5.7   Smailholm Scottish Anthrax

A very detailed account of the Smailholm Anthrax incident and the related incidents in Northumbria is presented in [Riley A, 2007]. A précis of the report is presented below.

On 8th July 2006 a 50-year-old man from the Scottish Borders died from septicaemia due to inhalation anthrax infection. Further testing of a house and the Village Hall where he had attended drumming classes also found evidence of widespread contamination by B. anthracis. Decontamination of some of the properties, notably the village hall in the village of Smailholm, was recommended by the Incident Control team.

Following initial tests to identify the extent of contamination, an expert panel was asked to comment on the extent to which the hall complex should be treated as contaminated. There were differing views from panel members ranging from only the specific sites sampled and found to be positive should be considered as the contaminated areas, to the entire hall complex should be considered as contaminated.

Similarly there was no single consensus view on the preferred decontamination method or agent. The final choice of decontamination agent was guided by best evidence in similar situations, in this case the NAS report referenced in Section 5.1 of this report.

Successful decontamination, principally using Chlorine Dioxide gas, took place in March 2007 over a 10 day period.

Riley (2007) notes that "Throughout the incident, communications with involved families, the public, the media, local & national partner agencies and politicians were specifically considered at each Incident Control Team meeting.

Specific efforts were made to liaise closely with family members directly involved. Channels of direct communication were opened and continuity of contact was considered important with rapport established and maintained whenever possible.... After liaison with the Village Hall Committee, it was decided that local members of the community preferred to be informed via the committee rather than by direct mail drop from NHS Borders. These wishes were respected and direct liaison with the Village Hall Committee was established for the remainder of the incident. Local elected members were briefed of decisions taken."

The team in charge of the incident also prepared a series of communication materials throughout the incident including relevant Q&A sheets.

A key principle of the incident communication strategy was to agree that all information to be released should be channelled through one central point and to achieve this, a communications group was established, with representation from all key agencies involved.

Although there was much media attention to the incident at the time of its discovery and there was undoubtedly some effort to generate a heightened level of interest and excitement among the public the incident was characterised by the relative absence of hype or panic. There are of course numerous factors surrounding the incident that will have contributed to this but perhaps the following three are relevant; i) There was a good deal of sympathy for the man who died from the contamination, essentially through no fault of his own, ii) the community in Smailholm is small and perhaps more closely knit than might otherwise have been the case and iii) the incident team took great pains to ensure that Public Stakeholders were engaged and involved with decisions.

5.8   The Tokyo Metro Sarin Attacks

Although the doomsday cult Aum Shinrikyo is believed to be responsible for several attacks in Tokyo, the most well-known chemical terrorist attack occurred on 20 March 1995. The cult's motive was to stop police investigations against the cult's activities by attacking the subway station closest to the Tokyo Metropolitan Police headquarters. The attack occurred during the morning rush hours. The sect members left eleven bags in five subway wagons in three different subway lines all heading for the same station. The plastic bags were punctured using umbrellas. The bags contained about 6 litres of impure sarin (30%).   The terrorist attack resulted in 12 deaths and thousands poisoned and seeking medical care (See D2.1).

Following the attack, the Japanese authorities decontaminated the affected rail stock using weak bleach solution starting about 8 hours after the attack. The subway was back in service later on the day of the attack [Pangi, R].

Several authors have commented upon how little disruption to subway use was caused by the attacks and among other factors have concluded that the role of previous events, specifically the Kobe earthquake two months prior, limited transportation alternatives, the limited service disruption and the lack of damage to subway infrastructure may have mitigated the social amplification of the terrorist attacks (see [Praget F et al] for example).

It is also perhaps worthy of note that the speed with which the decontamination was undertaken may have also acted to reduce risk perception from residual contamination and perhaps the short timescales did not allow those who might be concerned about the incident to gather knowledge and form perceptions of risk, although this could also easily have been a negative impact.

Some observers have noted that there were cultural dimensions to the public response in so far as the Japanese population were used to following government guidance and perhaps had a high level of trust in the officials.   Others have noted that that situation may now have changed following the population's experience following Fukushima and some of the confusion regarding data released following that event.

It is also noted that the Sarin would have naturally degraded in the subway with exposure to heat and humidity but it is not clear if that information was made available to the public users of the metro system.

5.9   Illegal Drug Laboratory

Following a Police raid on a flat in London in the United Kingdom[13] (U.K) it was discovered that it had been used for the illegal manufacture of Methamphetamines.

---

[13] The location of the flat is confidential. The authors have first hand experience of these works.

Because of concerns regarding health risks, the Local Authority had prohibited access to the flat to anyone not explicitly authorised by them, including the owners. Ultimately, specialist contractors were permitted access to the premises, after a lengthy delay whilst Police forensic investigations were completed and prosecutions were sought. A preliminary survey was undertaken to provide costs and proposals for the decontamination works.

As this was the first instance of such a decontamination exercise in the U.K, the agreement of the decontamination end-point to be used proved somewhat difficult. In the first instance, the Local Authority proposed a surface contamination limit equivalent to the background contamination of bank notes in the U.K. for the same drug family. The decontamination contractors and the owners however, proposed that a more meaningful and less restrictive risk based standard should be adopted. After involvement of the Health Protection Agency a standard was agreed which was equivalent to that adopted in the United States, which was endorsed by the Colorado Department of Public Health and Environment among others.

At the same time, the owners of the flat were in negotiation with their insurers to recover the costs for remediation. The insurers argued that the contamination did not constitute "damage" to the property and so the costs were not recoverable under that part of the insurance. The insurers did agree however that the prohibition of access by the Local Authority had led to a loss of rental income and agreed to cover those costs.

The flat was subsequently decontaminated using a mixture of HEPA filtered vacuuming, repeated washing with water and detergent, removal and disposal to landfill of furnishings and also by removal and disposal by hazardous waste incineration of some items. A detailed waste management plan was agreed which listed all items and all surfaces in the flat and prescribed a decontamination method for each.

Post decontamination, surface samples were obtained by the contractor and were analysed by an approved laboratory. These demonstrated that the residual contamination levels, measured as methamphetamine HCl, were below the agreed end-point, but not that there was no residual contamination. The samples also showed trace amounts of other drugs such as Heroin and Cocaine.

The Local Authority undertook some independent confirmatory sampling and the premises were eventually released back to the owner for redecoration and re-use.

Due to the nature of the works and the associated criminal prosecutions, they were undertaken in confidence, whilst the remainder of the building remained occupied.

5.10 Summary of Conclusions from the Case Studies

The foregoing case studies highlight that there are a number of linked technical and psychological factors that can have a significant impact upon the acceptability of any decontamination exercise and the residual risk from contamination. These are itemised briefly below and are recurring themes throughout the rest of this document

- Acceptance of risk by the Public is a complex issues that depends as much upon the social and psychological aspects as the technical ones. It is a values concept not a technical one.

- Different Stakeholders use different criteria to assess the acceptability or risk.

- Technical decisions regarding the suitability of particular decontamination techniques and their application should not be made without Public input and involvement nor without the involvement of those responsible for undertaking the decontamination or for financing it.

- Decontamination techniques can range from the complex to the very simple. Stakeholders do not necessarily expect complex or highly technical processes.

- Communication with Stakeholders is an essential element of a successful decontamination project. This communication must commence prior to any incident as this helps to engender trust, which is an important commodity in the recovery phase.

- It is possible, through good stakeholder communication and involvement to reach a consensus on appropriate decontamination techniques and residual risk.

- Clearly defined lines of responsibility and authority and their links to Stakeholders are essential for acceptance of any planned actions.

- Financial considerations, such as those associated with Insurance can be significant following a contamination incident.

6.   European Insurance Arrangements and Agreements

A review of some of the European practices with regards to the provision of terrorism Insurance is provided in [Erwann M-K et al] and [Erwann M-K]. A review of the current provisions made by some leading European Companies for property damage and business interruption insurance is provided in [AIRMIC]. Contamination from other incidents, not within the realms of terrorism, is generally covered by insurance arrangements but terrorism is often considered to be a special case or is excluded from standard cover such that it warrants special discussion here.

.Some key examples of terrorism Insurance across Europe are provided below to demonstrate the varying approaches taken and to highlight some of the difficulties that may arise post incident. The consequences of a terrorist attack may not be in the direct form of Physical Damage and Business Interruption loss, but also indirectly via Denial of Access, Supply Chain Disruption, Loss of Attraction and Reputational Damage, but these latter aspects are not directly considered in this report, although they may be mitigated by the Protocols provided.

Notwithstanding the arrangements noted below, a recent review of the role of insurance following disasters, presented by [Baylis, M], notes that it is nevertheless necessary for the insured and the insurers and their advisers to ensure that they have a thorough understanding of each others businesses and the practices that might be adopted following such an incident.

6.1 Spain

Perhaps because of its history of Terrorist Incidents, Spain has formed the "Consortio de Compensation de Seguros" which as part of a state-backed insurance compensation scheme for "extraordinary risks" (including also storms, floods, earthquakes, riots), provides coverage for these risks in an add-on to property insurance. The policies are sold by private insurers but they are not financially responsible for the losses from such risks. The Consortio has paid claims following several Terrorist attacks in Spain but has never called upon the State guarantee which underwrites the scheme. The scheme is compulsory.

6.2 United Kingdom

The United Kingdom has an established mutual reinsurance organization (Pool Re) for commercial property and business interruption to cover claims only from acts of terrorism. Pool Re acts as a reinsurer for all insurers that wish to be a member of the pool. The U.K. Treasury provides Pool Re with unlimited debt issuance. Pool Re pays 10% of its collected premiums to the British government in return for this possibility to borrow government money. The scheme is voluntary.

6.3 France

GAREAT (Gestion de l'Assurance et de la RÉassurance contre les Attentats) was signed on December 2001 and launched on January 1, 2002. It is a co-reinsurance pool which includes all insurance companies licensed to operate in France. Insurance against terrorism is compulsory in France, so that all firms are covered against terrorist action on French territory. Insurers may then reinsure with the pool subject to the risk meeting certain criteria. The risk sharing under GAREAT is organized under several layers. Ultimately the top layer is an unlimited guarantee above 2 billion euros provided by the French government via the Caisse Centrale de Réassurance, a state-owned reinsurance company. The scheme is compulsory.

6.4 Germany

Extremus AG was created in November 2002. It is a private insurance company dedicated to covering terrorism risks and composed of 15 insurers and reinsurers domiciled in Germany. As is the case in France, Extremus also benefits from property reinsurance from the federal government. The government provides 8 billion euros of reinsurance in excess of 2 billion covered by the private market. In exchange, the government receives about 12.5% of the premiums collected by Extremus. The scheme is voluntary.

6.5 Elsewhere in Europe

A similar variety of schemes apply elsewhere in Europe, with for example, Denmark, Finland, Italy, Norway, Portugal, Sweden and Switzerland not having any state backed schemes and terrorism insurance being optional, whereas in Belgium there is no state backed scheme but insurance is compulsory [Willis].

6.6 Pre-determined Arrangements

Regardless of the arrangements that prevail within any member state or elsewhere it is becoming common practice to discuss residual decontamination measures with insurers. Through this process the Organisation gains a clear idea of the financial remuneration available for residual contamination clean-up. The Organisation should be clear about the financial situation before starting the recovery operations and most certainly before they are complete, although the Litvinenko incident discussed in Section 5.3 and the case discussed in Section 5.9 (Illegal drug manufacture) highlighted instances where these discussions had not been held and the definitions of damage were ambiguous at the outset.

6.7 Contractor Framework Agreements

Government authorities with responsibility for CBRN incidents may have prior agreements with contractors for CBRN incidents, including recovery operations; in the United Kingdom this had been established under a government agency known as the Government Decontamination Service [GDS]. These are normally called 'Framework Agreements' and set out the arrangements for, among other things, clean-up. The Agreement includes costs for personnel and equipment that are agreed in advance (but with a variation of costs clause) of any incident. This arrangement enables authorities to proceed with the incident without any contractual delay. Such prior financial arrangements are also of benefit in any insurance claim. The author's are not aware of any further instances within the EU where these arrangements have been formally established to the same degree.

PRACTICE Deliverable D7.3 will provide a procurement protocol for acquiring the necessary equipment and training for preparing for a CB incident.

7. Applicability of standard Decontamination Practices to Incidents and Attacks

The examples quoted in Section 5 clearly demonstrate that the elements noted in Box 1 are not always present in the event of an unplanned incident or terrorist attack or at least that they are uncertain. The application of normal decontamination practice in such situations is therefore not appropriate without modification. Not least amongst the issues that are relevant to this document are those associated with the Official bodies (Local Authorities and the like) assuming overall control during the recovery phase and the degree to which this can cause difficulties for an Organisation.

Thus, while linear approaches are successful for normal settings, they may not provide sufficient flexibility to ensure acceptable decontamination following terrorist incidents or accidents. The studies that have been presented show that the key factors that need to be considered include the understanding of risk perception, risk acceptance and stakeholder management. The studies also emphasise the dynamic nature of post incident decontamination both in respect of the in-house management practices and with respect to the interaction with Stakeholders. The implications of this dynamic behaviour are considered in Section 7.4 (Dynamic Stakeholders).

These topics are the subject of the following Sections which then lead into a summary of the themes to be developed in the Protocols presented in Part B.

7.1   Stakeholders and Perceptions

From the brief reviews presented in Section 5 and the referenced detailed studies of previous incidents (see for example [Clarke, L] and [NAS]) it is clear that the process of achieving consent for decontamination standards is complex and that it must have a sociological element. However, what is also apparent and is perhaps obvious with hindsight, is that the process of ensuring an acceptable conclusion needs to start before a contamination incident has occurred. This means that, among other things, the Organisation's culture needs to be aligned to the idea that there could and probably will be a large range of stakeholders involved post incident, each with slightly different priorities and goals, each of which the Organisation has committed to satisfying. Furthermore, the studies show that communication **to** these stakeholders is not the key but rather communication **with** and **involvement of them** in the project at an early stage.

For some of these Stakeholders (e.g the Local Authority or their equivalents) the Organisation will be obliged by legal enforcement to satisfy them (as to the level of decontamination and cleanliness) and for some they will be obliged by commercial arrangements (e.g. insurers); as such these do not represent a new dimension. Others (like user groups, neighbours, visitors etc) may only become important or even come into existence after an incident. The cultural change required here is to recognise that although these other groups do not affect day to day operations they are nevertheless important to the process of obtaining acceptance of risk following an incident. This applies where risk actually extends to them or is perceived by them to extend to them.

Three fundamental groups appear to exist which we have termed the "Organisation", the "Official" and the "Public Group". The Organisation represents the owner/operator or guardian of the property or assets that have been negatively affected by the incident. They have an imperative to decontaminate their assets which may exceed that arising from external pressure from government bodies, the press, citizens and users for example (see Box 2).

```
┌─────────────────────────────────────────────────────────────────────────┐
│                     Box 2: Examples of Stakeholders                       │
│                                                                           │
│   •   People directly affected                                            │
│                                                                           │
│   •   Local communities                                                   │
│                                                                           │
│   •   Decision makers                                                     │
│                                                                           │
│   •   Relevant interest/activist groups                                   │
│                                                                           │
│   •   Landowners / Landlords                                              │
│                                                                           │
│   •   Neighbouring industry / businesses                                  │
│                                                                           │
│   •   Schools, nursing homes, etc.                                        │
│                                                                           │
│   •   Local leaders, councillors and local MPs                            │
│                                                                           │
│   •   Local authority and regulatory agency contacts,                     │
│                                                                           │
│   •   Decision makers and financial staff within the local authority or   │
│       government                                                          │
│                                                                           │
│   •   Other regulatory agencies                                           │
│                                                                           │
│   •   Developers                                                          │
│                                                                           │
│   •   Conservation bodies                                                 │
│                                                                           │
│   •   Legal and insurance advisors                                        │
│                                                                           │
│   •   Local health trust(s) or equivalent, and the local public health    │
│       agency                                                              │
│                                                                           │
│   •   Local media                                                         │
└─────────────────────────────────────────────────────────────────────────┘
```

This also includes employees of the Organisation who play a role in the control or management of an incident. The Official group represents those with a mandate and authority granted to them under legislation or other national or local government arrangements.

The Public Group represents those interested in or affected by an incident that does not fall into one of the other two groups. There is a degree of overlap between memberships of these groups as shown in Figure 6.

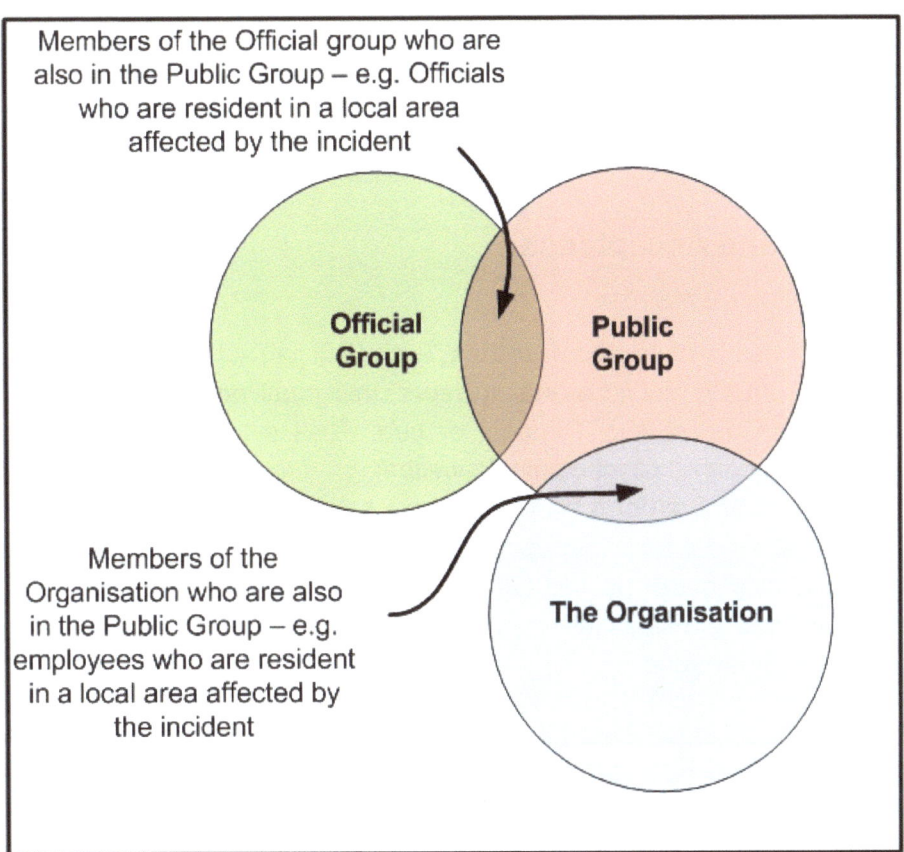

Members of the Official group who are also in the Public Group – e.g. Officials who are resident in a local area affected by the incident

Official Group

Public Group

The Organisation

Members of the Organisation who are also in the Public Group – e.g. employees who are resident in a local area affected by the incident

Figure 6: Three Stakeholder Group Membership

During an Organisation's normal range of operations the overlaps are meaningless as the Public Group does not exist because there is no group of people at risk outside of the Organisation's every day control. In the event of an incident which leads to the potential for exposure of a larger group – outside of the Organisation's normal controls - then it is possible that some employees of the Organisation or of the Official Group will also be in the Public Group. These stakeholders will be influenced by some of the same concerns and perceptions of the Public. In short, in these overlaps there may be conflicts of interest.

There is an added complication to the picture in Figure 6 in terms of goals, which is perhaps best illustrated by comparing specific examples; a Local Authority (an example of the Official group) will largely have as its goal "to ensure the safety of the public", which at first sight would appear to be the same as the goal of the Public Group. In so far as the Public group behaves as a group this is true, but the members of the Public Group are not necessarily bound together in as strict a hierarchy as the Official Group and its members and as such they may apply a more personal bias to their goals; their goal may be to "ensure my safety". This is rather more subjective as it is based upon or biased by perception, trust, personal circumstances and experience, for example.

The other clear message is that the tools of the Official Groups and the like are Risk Assessment, Expert Opinion and Precedence for these at least provide security against legal action that these bodies require, in addition to providing them with the confidence that they have discharged their responsibilities for public safety. These are therefore outwardly at least, largely fact based organisations. However, while the Public Groups may also place some reliance on these factual assessments, they are not constrained to them and are free to balance them with a more perceptual basis of assessment. The Public Groups have a social or sociological context to their decision making. This is linked to the perception of risk as discussed in the following section.

7.2   Risk Perception and Acceptance

The traditional definition of risk is often expressed as "Risk (Official) = Likelihood of occurrence x Magnitude of damage"[14],[15]. In the Public arena however, this must be modified by a third term which represents some measure of social impact or peril. Risk in the Public Group may well extend beyond physical health; psychological wellbeing, family health, friendships and communities are all exposed and all form part of the risk perception [Clark, Klinke and Ortwin]. Furthermore it may be that the Public Group are those that will have to bear the negative consequences of public policy made by the Official Group[16]. Given the often numerical and empirical nature of the 'science' of risk assessment it is not surprising that the sociological aspects are often overlooked or simply avoided.

If we call this extra term the Social Impact Factor then the way that the Public Group measure risk may be represented as

Risk (Public) = Likelihood of occurrence x Magnitude of damage x Social Impact Factor

But the Social Impact depends upon other factors such as the nature of the Threat and its Hazards (e.g. is it a chemical a biological or radiation hazard?), how close to home it was, whether it has happened before etc. Some measure of Threat and Hazard characteristics is therefore also required. Thus, the simple equation is not really adequate to represent the true nature of risk as perceived by the Public Group which may not be a simple first order equation (see Table 2.1 in [NAS 1989] for example and Appendix E in [NAS 2011]). It is argued by others [Klinke and Ortwinn] that this separation of variables nevertheless results in a useful model.

The difference is perhaps best summarised by the following

• Official Groups use fact based Risk Assessment to select and justify goals aimed at ensuring safety.

• Public Groups use Risk Perception, which may be advised by Risk Assessment, among others factors, to determine if they agree with or accept the Official standpoint.

It is perhaps also not surprising that an Official Group who traditionally assess risk to the Public Group via one method sometimes come up against resistance from that group, who have reviewed the same risk using a different method or who have considered other risks from the

---

[14] Traditionally, within the scientific community, damage to human health and the environment are at the fore of risk analysis and risk management.

[15] Sometimes (e.g. [Cox]) this is expressed as Risk=Threat x Vulnerability x Consequence, but this does not affect the discussion presented here.

[16] This perhaps also implies that there should be some measure of sustainability in the equation too. This is also argued for in [Subr:im].

Threat. The Public Group necessarily disagree with the methods used by the Official Group to assess the risk to them, rather they do not agree that the risk has been fully assessed or that the definition of risk is correct. These subjective aspects hardly come into play in the Official assessment process. One approach recommended by [Klinke and Ortwinn] is that criteria for evaluating risks should be developed from the social discourse about concerns, while the "objective" measurement should be performed by the most professional experts at hand.

Thus, what each group should be doing is seeking acceptance from the other of their measure of Risk and of their conclusions. Note that acceptance does not imply complete agreement; rather it implies some form of agreement to disagree on some aspects, but with overall agreement. This, in turn implies a need for trust and credibility between the parties [Harnett, Peters et al].

In other areas of risk management, such as that of contaminated land management, this call for a wider involvement of stakeholders has already been picked up. For example, [Ferguson et al 1998] notes that "...the use of the results of scientific risk assessment in environmental decision making must take the perception of various risks and other social issues into account. The development of coherent risk communication strategies is important":

In [Ferguson et al 1999] the authors note that "At the same time, however, risk management and policy would be overstrained if each risky activity required its own strategy of risk evaluation and management." Therefore what risk managers need is a broader conceptual model for management that ensures integration of the social considerations and the relevant multidisciplinary (scientific) approaches and, on the other hand, allows for established routines and easy-to-implement protocols. In asking others to participate more fully in the decision-making process, however, we may be calling on skills and techniques with which they have little or no experience [Teuber, A].

In summary, an "Organisation" may be caught between the two camps of risk/fact based requirements from the Official Group and the more precautionary approach of the Public Groups but has to satisfy both. The Official Group do not need to do this and perhaps may exit the incident before these broader questions are resolved – unless the pressure is great (e.g. from the press). Table 3 shows some of the types of factors that influence risk perception and which therefore need to be considered (see [Klinke and Ortwin], [Becker S.M], [SNIFFER] and [Tønnessen and Wesiaeth L] among others).

Table 3: Typical Dimensions of Risk Perception

Typical Dimensions of Risk Perception	
Magnitude or extent of damage or Catastrophic Potential	Inequity (between benefits of decontamination and those bearing the risks)
Probability of occurrence	Dread/Fear
Uncertainty	Familiarity / Knowlegde
Geographic extent	Choice / Voluntariness
Persistency/Reversibility	Trust
Effects on future generations	Media Attention
Latency	

7.3   Dynamic Risk Assessment

The dynamic nature of a decontamination exercise, where the exact nature of the contamination and hence the risks are not known until work commences are recognised in industry and typically a Dynamic Risk Assessment (DRA) approach is adopted – especially among emergency services (see [Fire Services Inspectorate] for example).   This is a continuous process of identifying risk and is applied immediately upon arriving at the scene of an incident and throughout the process of dealing with an incident. Figure 7 shows the DRA process. It is seen as the responsibility of all stakeholders involved at the scene of an incident to conduct independent DRAs on their own behalf and in concert with the emergency services and other stakeholders as necessary.

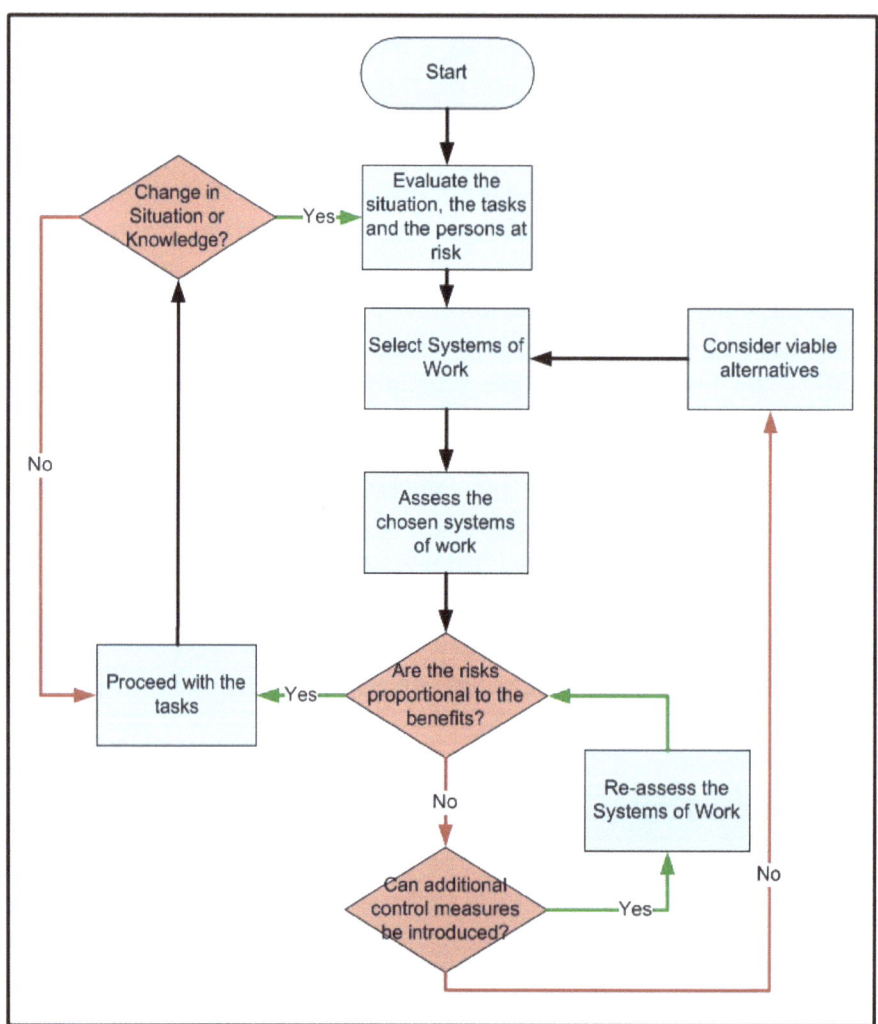

Figure 7: Dynamic Risk Assessment Process – Adapted from [Fire Services Inspectorate]

7.4   Dynamic Stakeholders

Consideration of (Linear Approach to Decontamination) shows that Stakeholder views are not gathered in the standard approach as they may be taken to have already been established through normal business activities and compliance with extant legislation.

The foregoing discussion not only shows the importance of stakeholder involvement but also the potential for stakeholder groups and their goals to differ and to be determined by both the type of the incident and its impact. i.e. an incident has implications not only for stakeholder membership but also for stakeholder objectives/views. Figure 8 shows that the nature of an incident can determine who will become stakeholders and what values they may apply to the incident and also how the response to an incident may further modify both of these. Thus stakeholder membership and goals are dynamic.

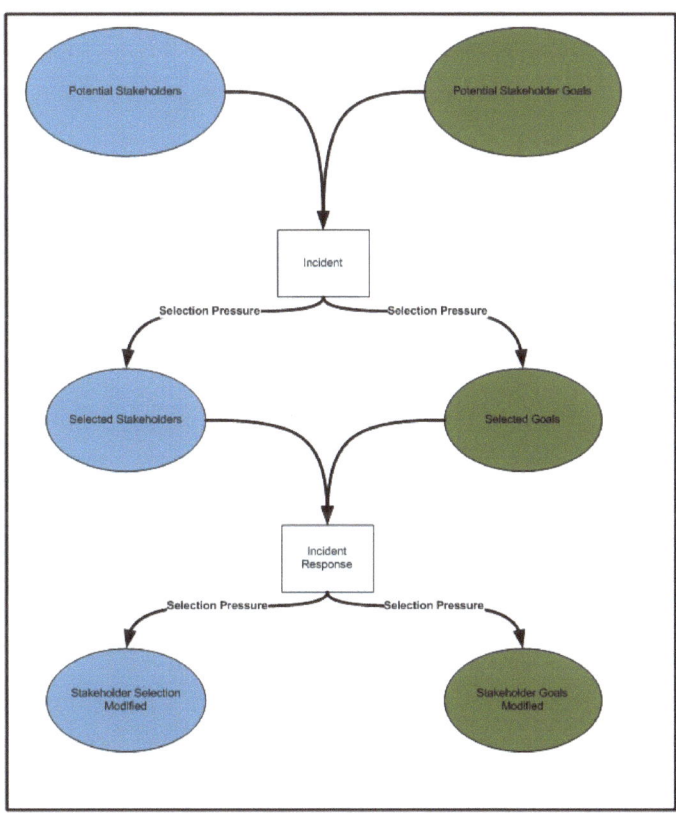

Figure 8: Selection of Stakeholders and their Goals

Furthermore, during normal operations, many Organisations only have to satisfy the members of the Official Group and thus standard operating practices and the like are orientated towards demonstrating compliance with the fact based risk assessment approach[17] and this becomes the norm of operations. It is also noteworthy that these compliance requirements generally change

---

[17] It is important to note here that these approaches do commonly include approaches that address Societal Risk, but this addresses aspects like the number of people affected or the number killed which is only a sub-set of the hypothesised Social Impact Factor (see Table 3).

slowly and with warning so that organisations do not have to respond quickly to changes and have time to take them on board in their organisational arrangements. To borrow Tuckman's paradigm [Smith], the "storming, formal, norming, performing" cycle (Figure 9) does not need to be re-enacted with any haste in normal operations. Specifically there is time for arrangements to mature and for the practitioners of those arrangements to become comfortable with them. Clearly, this may not be the case following an incident, where the time pressures and changing uncertainties (among other factors) can lead to frequent re-evaluations and little time for the "norming" phase. The pressures may lead to a need to completely bypass the norming phase and an attempt to go to the performing phase without having established the norms required to ensure effective performance – with obvious detrimental consequences.

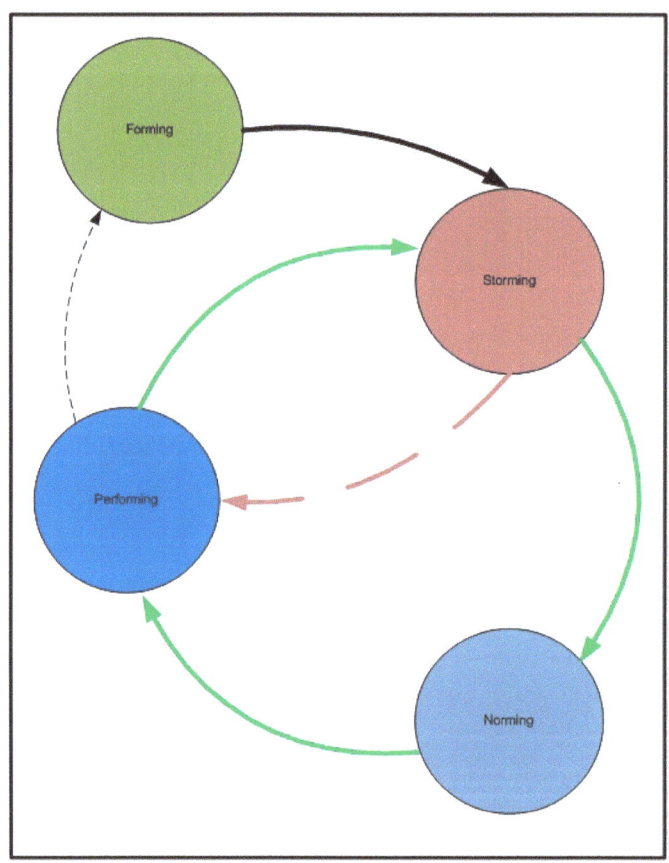

**Figure 9: Tuckman's Paradigm of Management and
the Effects of Crisis Management Bypassing the "Norming" Phase**

A further dynamic which needs to be accounted for and which is suggested by the foregoing and [Whicker et al] is that the level of knowledge and information available to stakeholders can change during the timescale of an incident and that this also effects the selection of their goals and objectives. Figure 8 can therefore be augmented as shown in Figure 10.

Strictly, the knowledge space does not necessarily have to grow, but rather it can change, but the net effect is the same. The question that clearly arises from Figure 10 is "where do the stakeholders get their information (and hence knowledge) from." It is not suggested here that the Organisation should control or try to control the information available to Stakeholders but rather the figure explains the need for an Organisation to understand and have access to the same sources of information as the other Stakeholders, so that they can understand and appropriately respond to that information, and also the need to have routes for communicating their own information to the Stakeholders.

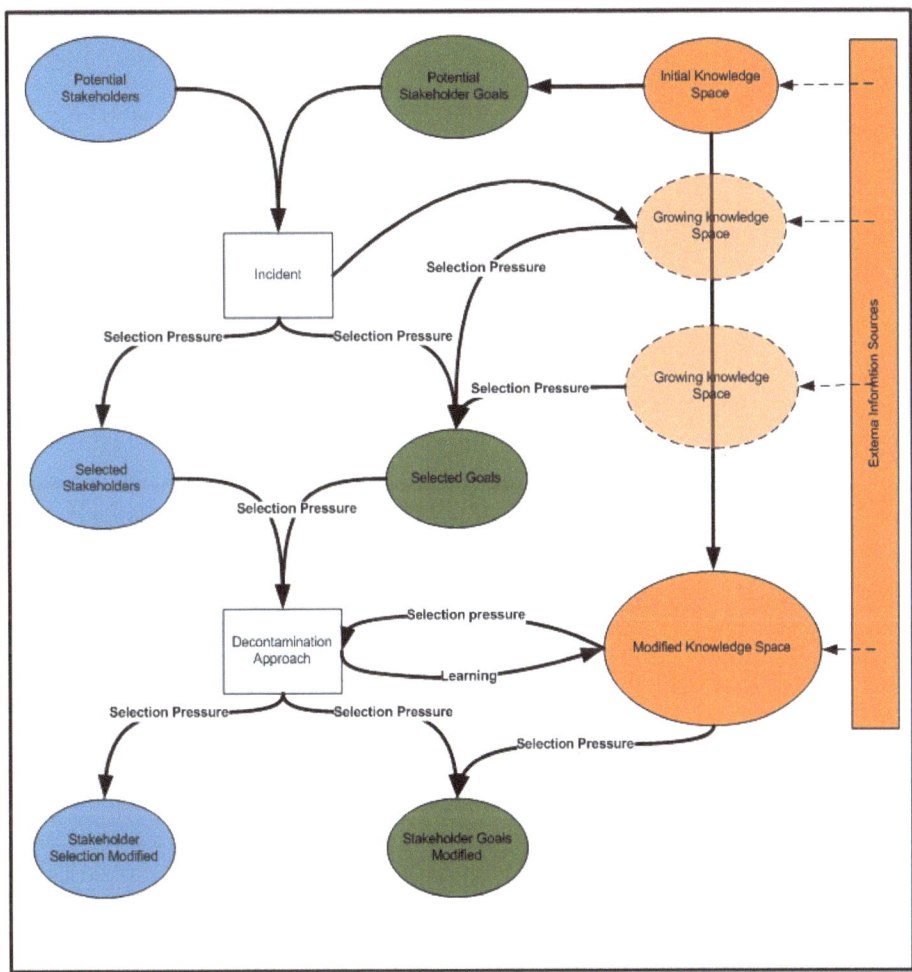

Figure 10: Effects of Knowledge Growth on Decontamination Strategies

Figure 10 illustrates the need for a good two-way process that sustains the trust of those affected and identifies actions needed to address their concerns. The case studies noted earlier also show that the nature of risk perception following an incident may require different communication strategies and mechanisms than those adopted during normal operations. Thus there is a requirement for risk communication training or at least preparedness. Subr:im (2007) provides semi-empirical evidence for the value of such preparedness.

A corollary to this, as shown in Figure 11 is the potential effect that Stakeholder perceptions may have on the extent of sampling and monitoring that is implied to confirm acceptable decontamination. Figure 11 shows that the heightened perception of risk that may arise following an incident can cause an increase in the amount of sampling and monitoring required to obtain stakeholder acceptance of the decontamination process and residual contamination levels (compare to Figure 5). Thus, justification of risk from residual contamination not only requires justification of the chosen standard but also justification of the chosen method of measuring that contamination to achieve the required statistical confidence.

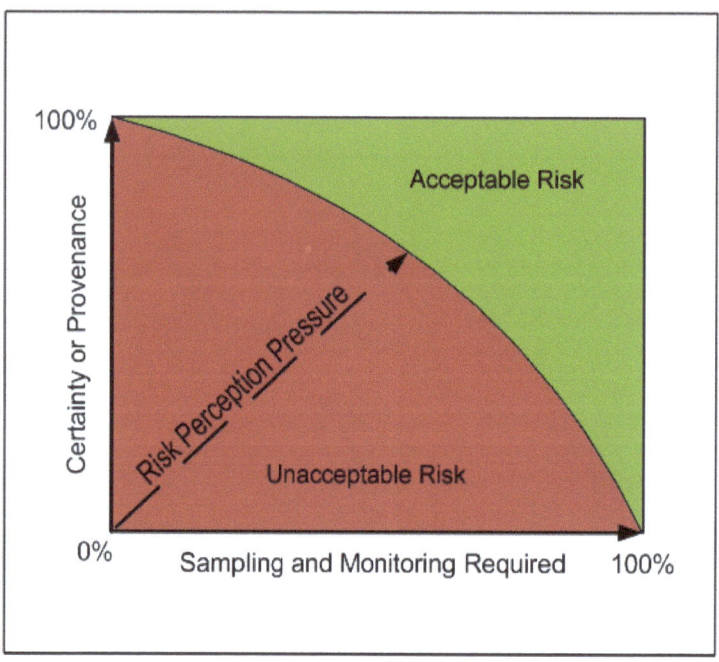

Figure 11: Revised Risk Acceptability, Monitoring and Provenance from Heightened perception

7.5   Stakeholder Responsibilities

In the same way that Stakeholder membership and goals vary as an incident progresses, it is clear that their responsibilities also change but that their authority may persist. For example, in the immediate aftermath of an incident it is the European norm that State bodies will assume control to ensure the protection of public and environmental health and that the immediate burden of costs will also be carried by the State. Once the immediate threat to public health has subsided and the incident enters the Recovery Phase (see) the Official bodies may still exercise authority over the remediation works but financial responsibility, in those countries where insurance cover is purchased by the Organisation, passes to them.

Thus, the decontamination standard may well be determined primarily by the Official Group and indeed it may be imposed on the Organisation via the equivalent of an enforcement notice, but it is the Organisation who will pay for implementing it and be responsible for achieving it.   In some countries, if these costs are not covered by specific state imposed insurances then the Organisation may have little control over the extent of these decontamination costs that are recoverable. However, the Organisation clearly has a role to play in ensuring acceptance of these decontamination standards by the stakeholders and in ensuring that the practicable limitations (in terms of financial constraints) are understood by the stakeholders.

There is a further complication in that an Official body may impose or at least recommend a decontamination standard – for their purposes – but under EU legislation it is the Organisation who is responsible for determining acceptable risk and for developing risk assessments to justify that risk. Whilst an Organisation may well be able to refer to the Official guidance in respect of justification for their risk assessment conclusions it is not clear that reliance upon them would constitute a discharge of the Organisation's legal duties to assess the risk. Thus the responsibility for determining end-points or decontamination standards also has an additional dimension following an incident that is outside of an Organisation's normal regime of operations.

7.6 Stakeholder Management and Communication

It is also clear that, following an incident of the type discussed in this report, there will be a need for some form of management of the Official, Organisation and Public Groups.

The management of all of the types of issues listed earlier in this report is broadly covered in the topic of Stakeholder Management (SM) which has been extensively researched and written on – see for example [Collier D et al, EC and AccountAbility] for overviews, guidance and listings of reference material. In the context of this report, it is seen that the term Stakeholders includes the Official, Organisation and Public Groups. This broader definition of stakeholders is different to that sometimes adopted but is necessary to ensure that the complexities and dynamics of a post incident organisation are captured.

The key element for Stakeholder Communication is that it needs to focus on the issues that recipients most need to understand (but do not know already) and that recipients are given the time, help and opportunity to understand [Morgan M.G et al (2002)]. This clearly implies the need for an intercourse between the Organisation and Stakeholders to ascertain what those needs are. The references outlined above provide templates and detailed guidance on this process. Following on from this identification of the stakeholder needs is the process of communication (or engagement) and this is often undertaken in accordance with a Stakeholder Communication Plan, for which many templates exist [see SPARC for example].

Stakeholder management and communication are key themes addressed by the Protocols in Part B of this report.

7.7 Trust and Credibility

In a similar way, it is clear that Trust and Credibility are also important issues during any remediation process where agreement is sought to some end-point.

Along with the analysis of stakeholders and stakeholder management the topic of trust and credibility has also been extensively written about – see [Cozma R], [Mitchel J.V] and [Peters R.G et al] for example – and is a key theme for the Protocols in Part B. Some related discussion on the perception of risk and factors which influence trust is presented in PRACTICE deliverable D8.8 (Lessons learned from existing research on [the role of human and social factors in] achieving resilience and preparedness to CBRN incidents). In summary, engendering trust and credibility involves the following:

i) Being open and showing that you care and are concerned for the Public [Peters, R.G]

ii)   Defying the stereotype (i.e. being different to what the public may perceive as a general image for your industry) [Peters R.G.]

iii)   Using Multiple sources (i.e. do not just rely on the official view and data but look for alternative sources to support it and present a rounder picture; "more richly sourced stories increase reader's trust" [Cozma R] and [Mitchel J.V] ; if Government has taken over, then procure independent verification to satisfy the public.

iv)   Noting that people under stress typically want to know that you care before they care about what you know [Covello]

v)   Noting that participation builds trust in the process, trust in the organisation and credibility [Collier, D]

7.8   Conclusions Regarding Decontamination Management

The foregoing discussion has shown that a linear management approach to decontamination management is not suited to the types of problem being considered here and that a more adaptive and stakeholder centric approach is required.

A number of alternative approaches to the management of contamination have been suggested and trialled. Notable among these is the Adaptive Management (AM) approach as set out by [Whicker et al] for example. AM has been broadly applied to environmental management with some success [Johnson, B.L] and there is an Open Standard for it [CMP]. There are many parallels between the social aspects of environmental management and those arising from contamination of an area (or environment) which suggest that the approach may have value here.

AM has up to now been used where there is uncertainty over the best management approach to be taken but there is no immediate threat to health. It holds that no single best policy can be selected but rather that a set of alternatives should be tracked to gain information regarding the best course of action. Moreover, the AM approach recognises from the outset that the situation is dynamic and that it may be characterised by uncertainties over data and knowledge (the 'Knowledge Space' referred to in Figure 10). It also recognises that as the knowledge space grows and changes the acceptable strategy may also change.   The approach therefore appears to be able to deal with some of the issues raised by this document.

Passive AM (PAM) involves implementing one management strategy at a time. The strategy is monitored to collect information about its efficacy and this is fed back into the management approach so that the system can be modified to optimise its performance against the current goals and objectives (noting that both of these can change as the project proceeds). The key aspect of AM is that this learning and feedback is seen as a necessary part of the process and that the associated change is beneficial rather than being seen as an indicator of failure. There are also parallels with the Plan Do Check Act model (and RADAR Approach) used in the European Framework of Quality Management [BQF] and the "Six Sigma" model [ISO].   The six stages of the AM loop can be framed as "Assess, Design, Implement, Monitor, Evaluate and Adjust" to use the Six Sigma terminology.

In the AM approach the management strategies adopted at any given time are sometimes referred to or seen as experiments (or interventions), which the user uses to help point them towards the optimal strategy or to test a hypothesis. It is not suggested in this document that it is appropriate to experiment in the difficult and dangerous situations that may exist following an incident. Once one recognises that the operational environment is dynamic, however, and that there are always uncertainties then the chosen strategy can be seen to be the best strategy to deliver the required results but that it is nevertheless an experiment until the results are available and the basis of the strategy is proven or otherwise. The adoption of the AM approach in the post incident situation recognises that things may not proceed as planned and it incorporates the learning from such occurrences into the planning process.

Although stakeholder views are garnered in the AM approach, it lacks formalism for the integration of those views and a methodology for evaluating that input. [Linkov, I et al] have suggested that where AM is linked with Multi Criteria Decision Analysis (MCDA) tools and concepts, it leads to a management system that allows structured, clear decisions to be made incorporating feedback and ranking of stakeholder inputs. MCDA entails measuring the performance or utility of something against a multitude of criteria, which do not necessarily share a common basis – e.g. the criteria may include measurable physical effects and more subjective qualitative concepts such as novelty or complexity. In its simplest form the method uses some form of weighed sum to obtain and overall measure from these contributing criteria.

In a review of the use of Multi-Attribute Decision Analysis (MADA a subset of MCDA) Tompkins E.L. (2003) has noted that MADA tools may contribute to the rationality of the decision making process, but they do not necessarily improve the 'quality' of the final decision made. It seems appropriate to extend this conclusion to MCDA too. This is exactly what is required in the management processes discussed in this paper i.e. a rational and mutually understood perception of risks (within all Stakeholder groups) that influences the final acceptability of the implemented processes. Tompkins also advocates the use of these tools to explore with stakeholders the potential impacts of different sets of preferences.

It is suggested in this report that a similar approach can be used to account for the dynamic and stakeholder aspects of post incident decontamination justification discussed earlier in this report. Furthermore, the principles of MCDA can be used to help formalise the assessment of the potentially diverse range of criteria (e.g. those in Table 3) that Stakeholders may adopt. Figure 12 (adapted from [Linkov et al]) shows how this combined approach can be mapped against the three stakeholder model shown in Figure 6 and how MCDA and Adaptive Management practice fit together in the overall scheme.

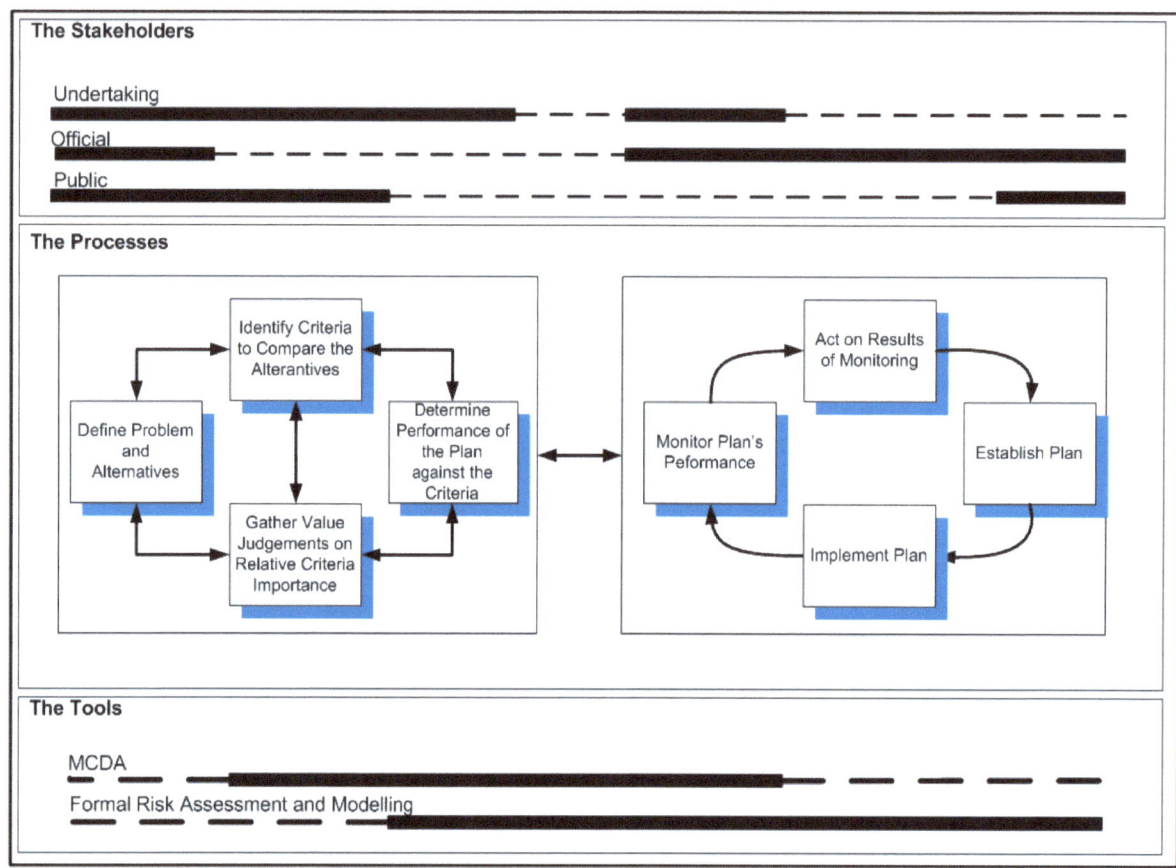

Figure 12: Combination of Adaptive Management and Multi Criteria Decision Analysis (adapted from Linkov et al).

Notes to Figure 12:

1. The involvement of the different stakeholders is shown by the solid lines (strongly involved) and dashed lines (lesser involvement).

2. The use of different tools is similarly shown by solid and dashed lines

3. The left hand side of the figure shows the process of setting the project goals, the determination of the multiple criteria that are to be used to measure performance and the gathering/processing of Stakeholder views on performance (all in consultation between the Official, the Organisation and the Public Group). MCDA is used to quantify the Public input and, to a lesser degree, the Official input and the Stakeholder's own views.

4. The right hand side of the figure shows the Plan-Do-Check-Act cycle, being fed by the output from the left hand side and vice-versa.

8. Conclusions

The foregoing discussions and observations from previous incidents have shown that the justification of risk from contamination is as much a sociological / organisational / cultural issue as it is a scientific one. Perception of risk rather than the actual risk can be paramount. Furthermore, it is not just the end-point that matters but also how it is arrived at. All regimes require the selection of an appropriate end-point with samples and monitoring data to provide the required degree of confidence that the end-point has been achieved. Difficulty arises from the definition of "appropriate" and "required degree" of remediation, or in the case of insurance arbitration, from the definition of what is "reasonable" in terms of cost. These factors encapsulate the "how clean is clean" conundrum. Moreover, the problem may not be a simple linear one - in so far as the actions taken following an incident may affect the acceptability of end-points and methods chosen at the outset. This is the additional social /psychological issue that must be addressed.

The key points raised in the main text are summarised below;

i)   Risk from residual contamination may be judged against different criteria by different stakeholders. Nevertheless it is possible to reach a **consensus** of understanding and an agreement on acceptability. This consensus can be aided by pre-agreements among stakeholders.

ii)   The acceptability of chosen decontamination standards is influenced by previous associations between Organisations and stakeholders, i.e. there is a dimension of trust involved in the acceptance and that trust can be earned from previous associations and agreements and carried forward into new circumstances. **Trust and credibility must be developed during normal operations and maintained during the post incident phase**[18].

iii)   Risk acceptability following an incident is a **dynamic** concept that requires careful management in the post incident phase and must be subject to frequent dynamic risk assessments.

iv)   Decontamination following an incident which involves stakeholders and issues outside of an Organisation's normal **practices must be managed to a different paradigm** than those that fall within the normal operations. Preparation and planning is key to ensuring successful response and recovery operations.

v)   Stakeholder Management is key to a successful decontamination; "stakeholders will have their own perspective on the restoration activities and goals, and extensive communication will have to be integrated for consensus on the best ways to proceed locally and globally toward the goal(s) of clean-up" [Whicker J.J et al].

vi)   For Stakeholder consent to play a role, it clearly cannot come into play only in the final stages of the decision-making process. Early involvement and engagement is key to obtaining future co-operation and trust. [Teuber A]. Consideration should be given to pre-arrangements and framework agreements.

---

[18] The corollary to this is that distrust can also be carried forward.

vii)   The principles and basic practices of Adaptive Management combined with MCDA may provide a valuable tool in the post incident decontamination phase as they may be better suited to the non-linear nature of these phases.

viii)   The acceptability of risk may have to be demonstrated and managed for a significant period post incident to a point where there is consensus among stakeholders that residual contamination no longer exists or it has been resolved to an acceptable level.

ix)   Organisations may be best represented in their discussions with Stakeholders by an independent party as this can help to engender trust and credibility.

The acceptability of risk that arises from residual contamination must therefore be judged as a balancing act where on the one side of the scales is the hazard arising from the contamination (as measured by objective scientific assessment) and on the other side is the cost and practicality of remediation modified to account for institutional and societal factors. Furthermore it must be recognised that acceptability is not necessarily a fixed time invariant concept but rather one that must be monitored and managed both during the decontamination phase and after it.

What the authorities need you to do to satisfy acceptable criteria and obtain authorisation to re-open may be less onerous than what the Organisation itself needs to do to satisfy their stakeholders (customers/clients and so on) and to re-gain their patronage.

9. Literature

Acton, J.M., Rogers, B and Zimmerman, P.D (2007),	*Beyond the Dirty Bomb: Re-thinking Radiological Terror,* Survival, vol. 49 no. 3
AccountAbility	*AA1000 Stakeholder Engagement Standard 2011,* AccountAbility (2008), www.accountability.org
AFColenco Ltd, Wilhelm Gottfried Leibniz Universität Hannover and DECOM a.s (2009),	*Dismantling Techniques, Decontamination Techniques, Dissemination of Best Practice, Experience and Know-how,* Co-ordination Network on Decommissioning of Nuclear Installations (CND) - contract no. 0508855 (FI60) from the European Commission's Research and Technological Development (RTD)
AIRMIC Research 2011	*Property Damage/Business Interruption (PD/BI) Insurance Benchmarking Report 2011,* AIRMIC Research.
Baylis, M (2012)	*Preparing for the Worst,* in AIRMIC News January 2012.
Becker, S.M. (2004)	*Emergency Communication and Information Issues in Terrorist Events Involving Radioactive Materials,* in Biosecurity and Bioterrorism: Biodefense Strategy, Practice, And Science Volume 2, Number 3, 2004.
British Quality Foundation (BQF) (2002)	*The Model In Practice, Using the EFQM Model to deliver continuous improvement,* British Quality Foundation, 32-34 Great Peter Street. London, UK.
Clarke, L (1989),	*Acceptable Risk? Making Decisions in a Toxic Environment,* University of California Press ISBN 0-520-07657-5
Clearance and Exemption Working Group (2005) (CEWG)	*Clearance and Exemption Principles, Processes and Practices for Use by the Nuclear Industry, A Nuclear Industry Code of Practice,* Clearance and Exemption Working Group
Collier D et al (2011), *SAFEGROUNDS:*	*Community stakeholder involvement (V3.0),* CIRIA, Classic House, 174–180 Old Street, London EC1V 9BP, UK.
Collier, D (2011)	*CIRIA W38 SAFEGROUNDS: Community stakeholder involvement,* , CIRIA, Classic House, 174–180 Old Street, London EC1V 9BP, UK
Covello, Vincent T. (2011)	*Risk Communication, Radiation, and Radiological Emergencies: Strategies, Tools, and Techniques,* Health Physics: November 2011 - Volume 101 - Issue 5.
Conservation Measures Partnership (2007) (CMP)	*Open Standards for the Practice of Conservation, The Conservation Measures Partnership, Version2.0 2007.*
Cox L.A, Jr (2008)	*Some Limitations of "Risk = Threat × Vulnerability ×*

	Consequence" for Risk Analysis of Terrorist Attacks, in Risk Analysis, Vol. 28, No. 6, 2008
Cozma, R (2006),	*Source Diversity Increases Credibility of Risk Stories*, Newspaper Research Journal • Vol. 27, No. 3 • Summer 2006
Department for Communities and Local Government (DCLG) 2009	*Multi-criteria analysis: a manual*, Department for Communities and Local Government, Eland House, Bressenden Place, London, SW1E 5DU
Dorison A. (2001),	*"The Industrial Disaster of Toulouse"*, Ecole des Mines d'Alès, September 2001. Dorison is the Director of the Ecole des Mines d'Alès, a French Institute for Engineering and Applied Science under the tutelage of the French Ministry in charge of Industry).
Erwann M-K (2011),	Report n°3: Financial protection of critical infrastructure, *http://www.institut.veolia.org/en/cahiers/protection-insurability-terrorism/ accessed December 2011*
Erwann M-K, Raschky P.A, (2011),	Report n°3: Financial protection of critical infrastructure, *The effects of government intervention on the market for corporate terrorism insurance*, in European Journal of Political Economy,
Esposito, M.P. Clarke, R. et al (1987),	*Decontamination Techniques for Buildings, Structures and Equipment*, Pollution Technology Review No. 142
European Commission (EC) (2009),	*Evaluating Socio Economic Development, SOURCEBOOK 2: Methods & Techniques: Stakeholder consultation*, http://ec.europa.eu/regional_policy/sources/docgener/evaluation/evalsed/sourcebooks/method_techniques/structuring_evaluations/stakeholders/index_en.htm, accessed November 2011.
Ferguson, C (1999),	Assessing *Risks from Contaminated Sites: Policy and Practice in 16 European Countries*, in Land Contamination & Reclamation, 7 (2), 1999, EPP Publications
Ferguson, C., Darmendrail, D., Freier, K., Jensen, B.K., Jensen, J., Kasamas, H., Urzelai, A. and Vegter, J. (editors) 1998,	*Risk Assessment for Contaminated Sites in Europe*, Volume 1. Scientific Basis. LQM Press, Nottingham.
Fire Service Inspectorate	*Fire Service Manuals Volume 2: Fire service Operations,* HM Fire Inspectorate, HM Fire Services Inspectorate,2009.
FOI (2011),	*Scenario template, existing CBRN scenarios and historical incidents*, practice.fp7security.eu

Government Decontamination Service (GDS)	*http://www.fera.defra.gov.uk/environment/governmentDecont aminationService/index.cfm/ accessed January 2012*
Harnett, T (2011)	*Consensus-Oriented Decision-Making (CODM),* New Society Publishers
IAEA (2009),	*SAFETY GUIDE No. WS-G-2.2: Decommissioning of Medical, Industrial and Research Facilities,* IAEA Safety Series Standards.
IAEA (2006).	*Management of Long Term Radiological Liabilities; Stewardship Challenges,* Technical reports series, ISSN 0074–1914;450, Vienna : International Atomic Energy Agency, 2006.
International Organization for Standardization (2011) (ISO)	*ISO 13053-1:2011*http://www.iso.org/iso/rss.xml?csnumber=52901&rss= detail *Quantitative methods in process improvement -- Six Sigma -- Part 1: DMAIC methodology,* International Organization for Standardization.
Johnson, B. L. (1999).	*Introduction to the special feature: adaptive management - scientifically sound, socially challenged?* Conservation Ecology 3(1): 10. [online] URL: http://www.consecol.org/vol3/iss1/art10/
Klinke A and Ortwinn R (2002),	*A New Approach to Risk Evaluation and Management: Risk Based, Precaution Based and Discourse Based Strategies,* in Risk Analysis Vol22, No. 6, 2002. Society for Risk Analysis.
Linkov I, Satterstrom F.K, Kiker G., Batchelor C., Bridges T., Ferguson E., (2006)	*From comparative risk assessment to multi-criteria decision analysis and adaptive management: Recent developments and applications,* In Environment International 32 (2006)
Mitchel, J,V (1992),	*Perception of Risk and Credibility at Toxic Sites,* Risk Analysis, Vol. 12, No. 1, 1992.
Morgan M.G, Fischoff B, Bostrom A, Atman, C.J	*Risk Communication A Mental Models Approach,* Cambridge University Press,2002.
National Academy of Sciences (NAS) (1989),	*Improving Risk Communication,* National Academies Press.
National Academy of Sciences (NAS) (2005),	*Re-opening Public Facilities after a Biological Attack – a Decision Making Framework,* National Academies Press.
National Academy of Sciences (NAS) (2011),	*Risk-Characterization Framework for Decision-Making at the Food and Drug Administration,* National Academies Press
NISD (Nuclear Industry Safety Directors Forum) (2005)	*Clearance and Exemption Principles, Processes and Practices for Use by the Nuclear Industry.*

Pangi R (2002),

Consequence *Management in the 1995 Sarin Attacks on the Japanese Subway System*, BCSIA Discussion Paper 2002-4, ESDP Discussion Paper ESDP-2002-01, John F. Kennedy School of Government, Harvard University, February 2002.

Peters R.G, Covello V.T, McCallum D.B,

The Determinants of Trust and Credibility in Environmental Risk Communication, in Risk Analysis. 1997; 17(1):43-54.

Prager F and Winterfeldt D,

Comparing behavioral responses to terrorist attacks on public transit systems: London, Madrid, and Tokyo, in Estimating behavioral changes for transportation modes after terrorist attacks in London, Madrid, and Tokyo, National Center for Risk and Economic Analysis of Terrorism Events (CREATE)

Scotland & Northern Ireland Forum for Environmental Research (SNIFFER)

Communicating understanding of contaminated land risks, Scotland & Northern Ireland Forum for Environmental Research (SNIFFER), Edinburgh Quay, 133 Fountainbridge, Edinburgh, EH3 9AG.

Smith, M. K. (2005),

Bruce W. Tuckman, forming, storming, norming and performing in groups, the encyclopaedia of informal education, www.infed.org/thinkers/tuckman.htm. Accessed October 2011.

Sports and Recreation New Zealand (SPARC)

Creating a Stakeholder Communication Plan, Sport New Zealand, PO Box 2251, Wellington 6140, New Zealand.

Subr:im (2007),

Communicating *Risk on Contaminated Sites: How Best to Engage with Local Residents*, Subr:im Bulletin SUB 06.

Teuber, A (1990)

Justifying Risk, in Proceedings of the American Academy of Arts and Sciences Volume 119 - Number 4, American Academy of Arts and Sciences, 136 Irving Street, Cambridge, MA 02138.

Tønnessen A and Weisaeth L

Terrorist events using radioactive materials: lessons for bioterrorism. In. Ursano, R.J., Norwood, A.E., and Fullerton, C.S. (Eds.) Bioterrorism: Psychological and Public Health Interventions, (pp. 165-99). UK: Cambridge University Press

Tompkins, E.L. (2003)

Using Stakeholders Preferences In Multi-Attribute Decision Makin:
Elicitation And Aggregation Issues, Centre for Social and Economic Research on the Global Environment, CSERGE working paper ECM 03-13.

Westminster City Council (WCC) (2007),

Framework strategy for dealing with radioactive contamination arising from the circumstances surrounding the death of Alexander Litvinenko, Westminster City Council.

Whicker J.J, Janecky D.R, Doerr T,B (2008) — *Adaptive Management: A Paradigm for Remediation of Public Facilities Following a Terrorist Attack*, in Risk Analysis, Vol. 28, No. 5, 2008.

Willis (2011) — *European Terrorism Exposure Bulletin, February 2011,* Property Investors Division, Willis Limited, Level 11, The Willis Building, 51 Lime Street, London, EC3M 7DQ, United Kingdom

1. Annex 1 : Summary of CBRNE Hazard Types

Term	Definition
Biological (B) threat compound	Biological threat compounds comprise micro-organisms, *i.e.* bacteria, rickettsia and viruses, and toxins, which cause disease in humans, animals or plants.
Chemical, Biological, Radiological and Nuclear (CBRN) incidents	CBRN incidents encompass all events in which exposure to C, B, or R threat compounds cause great harm to the health of people, animals and/or the environment, as well as incidents in which N materials undergoing fission cause harm through dispersed radioactive fission products or by direct irradiation. Such incidents may be caused by an accident or an intentional act.
Chemical (C) threat compound	Chemical threat compounds are chemicals that may pose a threat to humans or animals due to their toxic effects. **Note**. Numerous chemicals may pose a threat to humans, animals or the environment due to their toxicity, flammability or reactivity, or a combination of these properties. For the purpose of this project, C threat compounds are delimitated to those chemicals which pose a threat primarily due to their toxic effects.
Nuclear (N) material	Materials able to undergo fission, creating radioactive fission products and giving off direct radiation.
Radiological (R) threat compound	All radioactive substances can potentially be harmful if people are exposed. The determining factors are whether the exposure is internal or external, and the rate and duration of the irradiation.
Terrorism	The European Union's (EU) Council Framework Decision of 13 June 2002 on combating terrorism defines terrorism as intentional acts which "may seriously damage a country or an international organization" and are " committed with the aim of seriously intimidating a

Term	Definition
	population, or unduly compelling a Government or international organization to perform or abstain from performing any act, or seriously destabilizing or destroying the fundamental political, constitutional, economic or social structures of a country or an international organization" (EU Council Framework Decision, 2002). The same definition was used by FOI in a 2006-report evaluating crisis management capacity in the EU (FOI, 2006)
Toxic chemical	Any chemical which through its chemical action on life processes can cause death, temporary incapacitation or permanent harm to humans or animals. This includes all such chemicals, regardless of their origin or of their method of production, and regardless of whether they are produced in facilities, in munitions or elsewhere. (CWC, 1993).

The spectrum of toxic chemicals is wide and continues to expand. It spans from highly toxic chemical warfare agents, *i.e.* nerve- and blister agents, to toxic industrial chemicals, pharmaceuticals, bio-regulators and toxins. |
| Toxic Industrial Chemicals (TIC) | Toxic industrial chemicals (TIC) are industrial chemicals that are manufactured, stored, transported, and used throughout the world. |
| Toxin | Toxins are highly toxic chemicals produced by living organisms. The possible illegitimate use of toxins is covered by the prohibitions of both the CWC and the BTWC, thus toxins are, in principle, both biological and chemical threat compounds. However, it is most common to include toxins among the biological threat compounds due to their biological origin. |

PART B: Protocols for the Justification of Risk from Residual Contamination

1. Introduction to the Protocols

The degree to which an Organisation adopts or follows these protocols is a matter for their choice and will probably be influenced by factors such as perceived likelihood of being affected, funding, risk appetite etc. In any event, it is recommended that an Organisation records their reasons for adopting them or not – as per normal practice for risk assessment. The degree to which the Protocols may be developed by the Organisation may also be influenced by uncertainty over the types and extent of contamination to which they may be subjected. Nevertheless, it is the Author's belief that adoption of the measures set out in the protocols to the extent practicable now (however small that is) will yield benefits in the event of an incident. It will also help to prepare the Organisation's culture so that it is better able to deal with such incidents and will help to improve trust and credibility among Stakeholders.

Given the potential breadth of Organisations (in terms of size and interests) that may wish to adopt plans such as those promoted here, it is not possible to provide a one-size fits all approach. The Protocols are therefore, of necessity, specified at a high level. The protocols are not intended to be prescriptive but rather to highlight the key themes that need to be pursued by an Organisation in order that they might improve the likelihood of achieving a decontamination exercise that meets with both scientific and broader societal/stakeholder approval.

Where appropriate, the Protocols have been linked to existing European management standards such as those presented in EN ISO9001, EN ISO14001 and the European Framework for Quality Management. In this way it is hoped that Organisations will be able to implement the Protocols within existing frameworks of procedures and processes.

While the Protocols listed have been framed in terms of the requirements for the justification of risk following decontamination, post incident, they have implications for the broader issue of post incident management and an Organisation may choose to place them within that broader context within their management systems.

Before attempting to implement any of the Protocols the question that an Organisation must answer to itself is "how much risk am I prepared to take and how much effort am I prepared to take now to try to mitigate the remaining risk?".

2. Protocols

2.1 Protocol 1: Understanding Insurance

Title	P1: Understanding Insurance			
Relevant Sections of Part A	Section 6: European Insurance Arrangements and Agreements			
Purpose:	To ensure that insurance policy limitations with respect to recovery costs following a contamination incident are understood by those who are likely to be responsible for overseeing any decontamination works.			
Key Objectives:	i) Gain joint understanding (with insurers) of potentially conflicting definitions of end-points following decontamination process (See Figures 10 and 11). ii) To clearly understand extent of insurance cover available and exclusions that are applicable in the event of a contamination incident. iii) To ensure that insurers and paid advisors have a good understanding of the business before any actual event. [Baylis, M] iv) To get to know the insurers' procedures and how they would be likely to respond to a catastrophe. [Baylis, M] v) Understand arbitration procedures following a decontamination process.			
Suggested Processes:	Desktop study of example scenarios with Insurers / Loss adjusters to understand boundaries and issues. (See PRACTICE D2.1 [FOI] for Scenario Templates and examples of historical events). Based on the desktop scenario, fill in a claim form to help crystallise the issues.			
Output:	Memorandum of Understanding (or similar) between Organisation, Insurers and Stakeholder representatives (optional).			
Stakeholder input required?:	Stakeholder expectations garnered through Protocol 4 to be available prior to discussion with Insurers			
Useful References:	Scenario template, existing CBRN scenarios and historical incidents. PRACTICE WP2 deliverable D2.1. [FOI]			
Related Protocols	Protocol 2: Stakeholder Engagement			
Relevant Standards:	ISO19001	ISO14001	EFQM	BS25999
Clauses:		4.4.1	2. Policy and Strategy 4. Partnership and Resources	3.2.1.1, 4.4.2

2.2 Protocol 2: Stakeholder Identification and Analysis

Title	P2: Stakeholder Identification and Analysis
Relevant Sections of Part A	Section 7.6: Stakeholder Management and Communication Section 7.4: Dynamic Stakeholders
Purpose:	i) To identify key (potential) stakeholders prior to an incident. ii) To raise awareness of Stakeholder significance and breadth within the Organisation's organisation and culture.
Key Objectives:	i) Engage the Organisation's employees in the process of stakeholder identification and management. ii) Identify a list of potential stakeholders with whom the Organisation should commence early consultation (via the Stakeholder Management Plan [Protocol 3]). iii) Identify the potential Stakeholders prior to and following an incident. iv) Analyse and understand motivations for each Stakeholder. v) Ensure that the necessary resources are available to ensure implementation of the Stakeholder Management plan. vi) Establish Framework Agreements with potential decontamination contractors
Suggested Processes:	To identify stakeholders, start by asking for views from a wide range of parties within the Organisation (to help engender a culture of interest in their views) looking at the widest possible range of interested parties. These can include those with professional or technical expertise, those who are financially involved or impacted, and those with local and community knowledge or responsibility. Ask the following types of question [IAEA 2006]; i) Who has information and expertise that might be helpful following an incident? ii) Who has been involved or has wanted to be involved in similar risk situations before? iii) Who may be affected, with or without their knowledge, by the remediation / decontamination? iv) Who may be mobilized to act or angered if they are not included? Stakeholders may be both organizations and individuals but ultimately you must communicate with people so for each stakeholder there should be an identified point of contact - Organisations may wish to ask for communities to nominate representatives who can engage with them on behalf of the

Title	P2: Stakeholder Identification and Analysis

community.

Box 3 of the main text provides guidance on who may be Stakeholders. Stakeholders are those individuals or organisations who are likely to experience an impact, either directly or indirectly, as a result of the land contamination issue. Stakeholders are also those people who are able to influence whether or not a project will proceed. [SNIFFER]

It may be possible to prioritise or rank Stakeholders based upon their interest and authority but any such ranking is likely to be very dynamic and may require frequent re-visiting post incident. Similarly, influence maps can be constructed to show the relationships between stakeholders and their interactions, but these too are likely to be dynamic in the post incident phase. Nevertheless, the production of such rankings and diagram can be a useful starting point for incident planning.

Place Stakeholders into groupings of Organisation, Official and Public and note how each may move or overlap with other groups following an incident.

Output:	Stakeholder Analysis

Stakeholder input required?:	Implicitly

Useful References:	[SNIFFER] see http://www.sniffer.org.uk/project-search-results.aspx?searchterm=contaminated%20land&filterbycategoryid=9
	[Collier, D 2011]
	PRACTICE deliverable D7.3 is a Procurement Protocol for acquiring the necessary equipment and training for preparing for a CBRN event.

Related Protocols	Protocol 3: Stakeholder Management Plan
	Protocol 4: Stakeholder Communication Plan

Relevant Standards:	ISO19001	ISO14001	EFQM	BS25999
Clauses:	5.1a, 5.2, 5.5.1, 6.1b	4.3.2,	3. People, 8. Society Results	3.2.1.1, 3.3

2.3   Protocol 3: Stakeholder Management Plan

Title	P3: Stakeholder Management Plan
Relevant Sections of Part A	Section 7.3: Dynamic Risk Assessment Section 7.4: Dynamic Stakeholders Section 7.6: Stakeholder Management and Communication
Purpose:	i) To develop a plan for establishing stakeholder relations prior to an incident (to build trust and credibility). ii) To develop arrangements for communicating with Stakeholders following an incident. iii) To formalise the links between this plan and the Dynamic Risk Management Plan [Protocol 5].
Key Objectives:	i) Develop Stakeholder Management Plan ii) Execute the pre-incident sections of that plan and improve relationships iii) Ensure that the necessary resources are available to ensure implementation of this plan
Suggested Processes:	Having identified your stakeholders (see Stakeholder Identification Protocol 2), for each of them identify the following factors; your understanding of their interests, what their current overall position is (i.e. whether they are supportive, neutral or likely to be a blocker), what role you see them having (both now and following an incident), what actions are needed to ensure that they are engaged, how you will manage your stakeholder relationships (what agencies/mechanisms you will use). Note that the different Stakeholder Management Plans or different scales of plan might be required at different stages of an incident – see for example Figure 8 (Selection of Stakeholders and their Goals) and Figure 10 (Effects of Knowledge Growth on Decontamination Strategies) of Part A of this report. The plan must extend to post incident and post decontamination support. The plan must be known to all key personnel that are likely to be involved post incident and must be available post incident (i.e. not likely to itself have become contaminated). Give consideration to appointing a third party – possibly someone already respected within the community – to act as spokesperson or joint representative for the Organisation, as there is evidence to suggest that this engenders a greater degree of trust and credibility [Peters, R.G. et al].

Title	P3: Stakeholder Management Plan			
Output:	Stakeholder Management Plan and Annual Reports.			
Stakeholder input required?:	Implicitly			
Useful References:	[SNIFFER] see http://www.sniffer.org.uk/project-search-results.aspx?searchterm=contaminated%20land&filterbycategoryid=9 [Collier, D 2011]			
Related Protocols	Protocol 4: Stakeholder Communication Plan Protocol 2: Stakeholder Identification and Analysis			
Relevant Standards:	ISO19001	ISO14001	EFQM	BS25999
Clauses:	5.1e, 5.2, 5.5.1, 6.1b,	4.2g,	8. Society Results	4.2a, 4.3.3,

2.4   Protocol 4: Stakeholder Communication Plan

Title	P4: Stakeholder Communication Plan
Relevant Sections of Part A	Section 7.4: Dynamic Stakeholders Section 7.6: Stakeholder Management and Communication
Purpose:	To Define the methods and procedures to be used to communicate with stakeholders.
Key Objectives:	i) Develop a Communication Plan ii) Train nominated persons in the use of the plan and in risk communication, more generally. iii) Form formal links between the communication plans and the dynamic risk management plan (Protocol 5). iv) Understand the mechanisms by which Stakeholder Knowledge Space (for example, cognition, motivation) may vary – See Figure 8: Selection of Stakeholders and their Goals, in Part A. v) Test and maintain the Stakeholder Communication Plan. vi) Ensure that the necessary resources are available to ensure implementation of the Stakeholder Communication plan.
Suggested Processes:	Development of a Communication Plan that encompasses the following vii) The methods of communication to be used (channels, media etc – see Note 1 below) viii) The identification of those responsible for communicating (pre and post incident, as these may differ). ix) Messages and information to be provided and updated during normal operations (see Note 2 for suggested minimum information). x) A repertoire of audience tested, scientifically grounded pre-event messages that can be released almost immediately [Becker, S.M.] The plan should be tested and exercised under simulated conditions to verify its functionality – especially under the constraints that may exist post incident. The Plan must "survive the incident" – i.e. it must be in a form and/or location which will be available following foreseeable incidents.

Note 1: Stakeholder Engagement Channels

There are many potential methods of Stakeholder engagement / communication such as, Newsletters, Websites, Social Media (Facebook, Twitter etc), Public meetings / Surgeries, Workshops, Focus Groups etc. see Useful References for further guidance.

Notes 2: Recommended minimum information (adapted from [SAVESO and RAFAEL]

i)   Identification, by position held, of the person giving the information / responsible for communications.

ii) Simple details of how to contact the named person.

iii) An explanation in simple terms of the activity undertaken on the site.

iv) Adequate information on how the population concerned will be warned and kept informed in the event of an accident.

v)   Adequate information of the actions the population concerned should take, and on the behaviour they should adopt, in the event of an accident.

vi) References to any other emergency plans drawn up – especially those that relate to any off-site effects from an accident. This should include advice to co-operate with any instructions or requests from the emergency services at the time of an accident.

vii) Details of where further relevant information can be obtained (References to independently produced guidance and advice documents – e.g. National Guidance).

Note 3: Minimum Information Following an Incident

i)   General information relating to the nature of the incident, including its potential effect on the population and the environment.

ii) Information about what is known and not known in clear, simple language. The presence of uncertainty or variability should not be an obstacle to the provision of information to the public.

iii) Information about what people can do themselves, and what others are doing to address the problems

iv) The information process and what people can expect next (where, when, what medium, what they can do to access it etc).

v)   Methods for obtaining further information or for providing feedback.

Title	**P4: Stakeholder Communication Plan**			
Output	Stakeholder Communication Plan			
Stakeholder input required?:	Preferable			
Useful References:	[SNIFFER], see http://www.sniffer.org.uk/project-search-results.aspx?searchterm=contaminated%20land&filterbycategoryid=9 [SPARC] [AccountAbility]			
Related Protocols	Protocol 2: Stakeholder Identification and Analysis Protocol 3: Stakeholder Management Plan Protocol 5: Dynamic Risk Management Plan			
Relevant Standards:	ISO19001	ISO14001	EFQM	BS25999
Clauses:	5.1a, 5.2, 5.5.1,6.1b, , 6.2.2, 8.2.1, 8.4c,	4.4.7	5. Processes 8. society Results	3.2.1.1, 4.3.3.2, 4.3.3.3,

2.5   Protocol 5: Dynamic Risk Management Plan

Title	**P5: Dynamic Risk Management Plan**
Relevant Sections of Part A	Section 7.3: Dynamic Risk Assessment Section 7.4: Dynamic Stakeholders Section 7.6: Stakeholder Management and Communication
Purpose:	To develop a Risk Management Process that allows for semi quantitative involvement of stakeholder views and allows for dynamic feedback between stakeholder groups in situations that are unpredictable/ unforeseen.
Key Objectives:	To i) Gain joint understanding among stakeholders of the boundaries to decision making (financial, legal etc) that each group may be bound by. ii) Establish a risk management approach that is accepted by all Stakeholders in dynamic situations prior, during, and after the execution of an operational response. iii) Establish and agree among Stakeholders a set of metrics for the dynamic assessment of risk. Use Table 3 (Typical Dimensions of Risk Perception) of the main text as the starting point together with Figure 7 (Dynamic Risk Assessment Process). iv) Agree with Stakeholders a Multi-Criteria Decision Analysis (MCDA) technique to be used to 'quantify' the metrics (especially among the Public Group). Note: Many MCDA techniques exist and it may be necessary to obtain expert support in the selection and use of an appropriate technique. (See Useful References for further guidance and brief outline in Section 7.8 (Conclusions Regarding Decontamination Management) of the main text. v) Establish a risk management system within the Organisation which ensures that Stakeholder assent is obtained throughout the decontamination process and that feedback systems exist between the technical decontamination processes and the Stakeholder management processes. vi) Provide auditable records of the rationale behind decisions. vii) Ensure that the necessary resources are available to ensure implementation of this plan.
Suggested	Group study of example scenarios with Stakeholder groups to gain understanding of boundaries and issues and to test potential criteria and

Title	P5: Dynamic Risk Management Plan			
Processes:	methods for MCDA. Training to improve attitudes and behaviours to DRA. Formalisation of Dynamic Risk Management Processes to be applied to contamination incidents. Desktop study of example scenarios (in-house) to test Dynamic Risk Management procedures – using simulated injects from MCDA results.			
Output:	**Dynamic Risk Management Plan and Processes**			
Stakeholder input required?:	Implicitly			
Useful References:	[Linkov et al]. [CDLG] provides a comprehensive overview of Multi Criteria Analysis and also provides links to useful software tools.			
Related Protocols	All			
Relevant Standards:	ISO19001	ISO14001	EFQM	BS25999
Clauses:	5.5.1, 6.1b, 7.2.3, 8.1c, 8.2.1, 8.4c	4.2d, 4.3.2a, 4.4.7,	RADAR Model 4. Partnership and Resources	4.1.1, 4.2.1, 4.3.3, 4.4.2,

2.6 Protocol 6: Response Phase Action Plan

Title	P6: Response Phase Action Plan			
Relevant Sections of Part A	Section 7.0: Applicability of standard Decontamination Practices to Incidents and Attacks Section 7.6: Stakeholder Management and Communication			
Purpose:	To identify the actions and interfaces necessary with the Official stakeholders following an incident.			
Key Objectives:	i) To produce and agreed Action Plan to be followed by the Organisation and the Official Group during the Emergency Response Phase of an Incident such that information relevant to Recovery is available to the Organisation. ii) Agree methods of communication between the Organisation and Government Authority or other Official body mandated to take control of Incidents. iii) To jointly understand where information from the response phase is required by the Organisation for their Stakeholder Communication Plans.			
Suggested Processes:	Desktop study of example scenarios with Official Group to understand and agree methods of communication between responders and the Organisation. (See PRACTICE D2.1 [FOI] for Scenario Templates and examples of historical events).			
Output:	Memorandum of Understanding (or similar) between Organisation and Official Group An agreed Response Phase Plan			
Stakeholder input required?:	The Official Group			
Useful References:	PRACTICE Deliverable D5.11 will provide a Remediation Action Plan which will provide some further guidance in this area.			
Related Protocols	Protocol 2: Stakeholder Identification and Analysis Protocol 3: Stakeholder Management Plan Protocol 4: Stakeholder Communication Plan			
Relevant Standards:	ISO19001	ISO14001	EFQM	BS25999
Clauses:		4.3.2,, 4.4.7	4. Partnership and Resources	4.2